For Daniel, Oscar, Morgan and Felix

The Trees

The trees are coming into leaf
Like something almost being said;
The recent buds relax and spread,
Their greenness is a kind of grief.

Is it that they are born again
And we grow old? No, they die too.
Their yearly trick of looking new
Is written down in rings of grain.

Yet still the unresting castles thresh
In fullgrown thickness every May.
Last year is dead, they seem to say,
Begin afresh, afresh, afresh.

Philip Larkin, 1974

CONTENTS

IMAGE CREDITS   8

PREFACE   11

INTRODUCTION   15

CHAPTER ONE
Forest, Woodland or Plantation? Defining Tree-Covered Land   19

CHAPTER TWO
Forests Past, Forests Present: The Uncertain Narrative of Irish Forestry   29

CHAPTER THREE
Forest Protection: Local and Global   47

CHAPTER FOUR
Sustainable Forest Management   59

CHAPTER FIVE
Why Forests?   65

WITHDRAWN FROM DLR LIBRARIES STOCK

BAINTE DEN STOC

BAINTE DEN STOC

WITHDRAWN FROM DLR LIBRARIES STOCK

Why Forests? Why Wood?

Why Forests? Why Wood?

The Case for Multipurpose Forestry in an Uncertain Climate

Donal Magner

THE LILLIPUT PRESS

DUBLIN

First published in 2024 by
THE LILLIPUT PRESS
62-63 Sitric Road, Arbour Hill, Dublin 7, Ireland
www.lilliputpress.ie

Copyright © Donal Magner, 2024

10 9 8 7 6 5 4 3 2 1

ISBN 978 1 84351 891 4 hardback
ISBN 978 1 84351 883 9 paperback

Dedication page: "The Trees" by Philip Larkin, *High Windows*, Faber & Faber, 1974.
Quotations from: "VIII Once" by Eavan Boland, *Code*, Carcanet Press, 2013 (p. 15);
"Bog Oak" by Seamus Heaney, *Wintering Out*, Faber & Faber, 1972 (p. 29);
"The Scribe in the Woods" by Anon, *Medieval Irish Lyrics*, edited and translated by
James Carney, University of California Press, 1967 (p. 34); "The Ruin that Befell the
Great Families of Ireland" (c1720) by Aogán Ó Rathaille, *O Rathaille* translated by
Michael Hartnett, The Gallery Press, 1998 (p. 40); "Beech Tree" and "Poplar Memory"
by Patrick Kavanagh, reprinted from *Collected Poems*, edited by Antoinette Quinn
(Allen Lane, 2004), by kind permission of the Trustees of the Estate of the late
Katherine B. Kavanagh, through the Jonathan Williams Literary Agency (p. 185);
"Story" by Eavan Boland, *New Collected Poems*, Carcanet Press, 2005 (p. 219); and
Georgics by Virgil (29BC), translated by Peter Fallon, Oxford Press 2004 (p. 260).

All rights reserved. No part of this publication may be reproduced in any form
or by any means without the prior permission of the publisher.

A CIP record of this title is available from the British Library.

This publication has been part funded by The Department of Agriculture,
Food and the Marine. The author also acknowledges the support of
the Society of Irish Foresters.

Production: Magner Communications
Design: Grasshopper Graphics and Magner Communications
Printed by Grupo Grafo, Bilbao, Spain.

CHAPTER SIX
Why Wood?   115

CHAPTER SEVEN
The Right Trees in the Right Places for the Right Reasons   185

CHAPTER EIGHT
What Trees? Species That Adapt to Irish Soil, Site and Climatic Conditions   195

CHAPTER NINE
Our Future Forests   249

REFERENCES   269

GLOSSARY   277

INDEX   283

Sitka spruce, Gougane Barra, Co. Cork

IMAGE CREDITS

Photography is a major element in this book. In addition to my own library, I have drawn heavily from successive Wood Award Ireland catalogues which I have edited from 2014 to 2024 and a significant number of excellent images from Irish and international photographers which I gratefully acknowledge. These comprise (page numbers and positions – main (m); top (t); bottom (b)): Neville Dukes (2, 7-8, 26, 96, 186-187, 258); Eoghan Kavanagh, skyline.ie (6-7); Iain White, Fennell Photography (27); Adare Manor (41 b); Department of Lands (43); Coillte (44, 70 t, 162-163, 252, 261); Egor Vikhrev (46); Mike van Schoonder (52); Florian Rebmann (54), Donal Whelan (88-89, 224); Skyler Ewing (90); pixabay.com (91 t); John Thorne Photography (91 (b)); Wesley Atkinson (92); Kevin Collins (94); Frank Doyle Photography (95); Phil Mitchell (97 t); Odd Falch (97 b); John Magner (102); VESI Environmental Ltd. (107); Teagasc (108); Forest Service (108-109); Natural History Museum, London (117); Wikipedia (118 b); Museum of Vest-Telemark (120); University of Canterbury (122); Tony Corey (123); Built by Nature (128); Government of Ireland (130); Storo Enso (133); University of British Columbia, (135); Gisling (136); Donal Murphy (138 b); Ricardo Foto, Øystein Elgsaas (138); Christopher Hill Photographic/Alamy Stock Photo (140-141); Paul Tierney (140 t); Moreno Maggi (143 t); Renzo Piano Building Workshop (143 b); Sibelius Hall (144); National Opera House (145); Forest Industries Ireland (147); Cygnum Building Offsite (148); Noel Kenna (149 t, b); Medite SmartPly (134. 135); Andy Stagg (156-157); University of Galway (159); Michael Moran-Donal Murphy (160), Colm Kerr (161); National Gallery of Ireland (165); National Gallery of Ireland (164 t); Joseph Walsh Studio (164-165, 165 t, 166 m,t, 208-209); Mater Hospital (167); Alice Clancy (168-169, 177 m, t, 236-237); Alan Meredith (170, 171 b); Francis Morrin (171 t); Christian Richters (172); Seoi Sionna (173 b left); Naneen Project (174, 175);

Tony Corey (176 m, t); Cotton Coulson (176); Adrian McGreevy, Avondale EAK Ireland Treetop Walks (178, 180, 240); Barry Cronin (181 t); Jenni Roche (203, 211); Robert Bourke Architects, Dublin and Creatomus Solutions (181 b); Donn Holohan (182-183); Sean Lenihan (198); Donegal Co. Council (205); Noel Reddy (210); Knut Klimmek (215); Laura Fletcher (216 t); Elysia Taylor (217); Chaïm Factor (220 b); Tree Register of Ireland (221 t); Dennis Gilbert (222); Neill Warner (228); National Botanic Gardens, painting by Celia Harrison (229); Murray Timber Group (230); Shane Walsh (230-231); Aidan Monaghan (243); James Corbett Architects (244); The Wood Database (246); Dr Jim McAdam (263).The following remaining images are from my own collection: 10, 12, 19, 22, 23, 24-25, 30, 40, 41 t, 45, 54, 61, 62, 64, 69, 70 b, 71, 74, 81, 82, 85, 86, 87, 98, 103, 105, 106, 108 t, 109 t, 110-111, 118 t, 119, 121, 151, 172, 173 t, b left, 188, 192, 199, 200, 202, 203, 204, 206, 208, 211, 212, 213, 214, 216 b, 218, 219, 220 m, 221 b, 225, 227, 232-233, 235, 238, 241, 242, 245, 246-247, 262, 264-265.

Tables and figures are acknowledged on pages where they are set. In some instances, they are adapted by the author. Illustrations and paintings (and page number) by: Gustav Doré (20), in public domain; Brian Bourke, The Lilliput Press (28); Emil Shinkel and the Tree Council of Ireland (33); Manuscript St. Gallen, Stiftsbibliothek, Institutio de arte grammatical (20); Albrecht Durer, National Museum, Berlin (38); Forest Service, Woodland for Water (104); Woodspec (146); COFORD (150). In developing some tables and illustrations, my thanks to Gino Forte, Grasshopper Graphics for visualising my sometimes overly complex thought process. I'm referring in particular to figures and illustration in pages 31, 58, 68, 101, 126 and 127.

Scots pine and oak, Coronation Plantation, Co. Wicklow

Oak woodland, Poulanass or Poll an Easa (the hole of the waterfall), Glendalough Forest.

PREFACE

When I arrived in Glendalough Forest in 1972, one of my assignments as a State forester was to select trees suitable for electricity transmission poles. The forest was a hive of activity as tall, straight Douglas fir trees, were felled, extracted and prepared for the Electricity Supply Board (ESB) before they were hauled away and eventually erected to "carry the light" across the Irish landscape. Nearby, European larch and Scots pine were also used for transmission poles, and to supply local sawmills with logs for fencing, construction and pallets. Further up the mountain, Sitka spruce was harvested and delivered to sawmills throughout Ireland – north and south. As the seasons changed, so too did the ebb and flow of forest activity, from ground cultivation to planting, from pruning to tree felling and all operations in between. Looking back now, I realise how new all of this activity was and how reliant it was on State support. The State forest was less than 40 years old, while ESB's rural electrification scheme was little more than two decades in existence. Both activities resulted from a visionary approach by the State to invest in forestry and electricity when public funds were extremely limited at a time when access to European markets and investment was virtually non-existent. From 1961, Ireland's application for EEC membership had been repeatedly vetoed by French President Charles de Gaulle. Opposition ended with de Gaulle's shock resignation in 1969. Ironically, within two weeks of his resignation he visited Ireland on holiday, which included a visit to Glendalough. Forest staff recounted tales of de Gaulle's cavalcade speeding past into the deeply forested uplands, without stopping. Access to these upland forests was by a winding man-made road through a native oak woodland over the Upper Lake which along with the Lower Lake gave the valley its name: *Gleann dá loch* – glen of the two lakes. This short but steep journey through a beautiful forest and woodland inscape, presented a fascinating microcosm of Irish forestry. The contrast between the native woodland on the lower slopes and the upland mainly non-native coniferous was not only apparent in their natural physical differences in tree species, soils and ecosystems but in the way they were treated by us foresters.

The exotic conifers were regarded as the industrial forest workhorses while the native oak received deferential treatment, reserved for the elders of the forest. These had always been here, while the conifer interlopers were recent arrivals. This was the narrative as often presented and like all simplified histories it only tells part of the story. True, the oak was present here as climax forest as far back as 9000BC but like other broadleaves, it too disappeared as prints and paintings dating to the late eighteenth century depict a Glendalough landscape almost denuded of trees. While, today's conifers are introduced, the mountains were covered with the native conifer Scots pine as far back in time as oak.

The oak may be a poor imitation of the primeval forest in scale today, but its conservation and expansion are vital in restoring Ireland's woodland culture and our understanding of our arboreal prehistory. It should go hand in hand with the careful management of the introduced species. The species mix in Glendalough a half century ago demonstrated that a vibrant multipurpose forest culture values all trees highly, regardless of their origin.

Sitka spruce, Lugduff, Glendalough Forest, planted in 1933. One of the trees is 62m tall, currently the tallest tree in Ireland. The author was appointed State forester here in 1972 when this upland forest was managed commercially, while the oak (page 10) on the lower slopes was conserved as a native heritage woodland.

The oak woodland retains its permanency and unique ecosystem, while further up the hill the conifers are thriving on impoverished soils. They are now well into their second rotation, apart from a grove planted in 1933, which is preserved. Planted by some of the forest staff I met a half century ago, the grove contains a 62-m Sitka spruce, the tallest tree in Ireland. Maybe it's time we welcomed this and other exotics, which were introduced to Ireland in the early nineteenth century and began to make a major contribution to Irish forestry from the early 1900s. Wise forester Gerry Patterson, a strong advocate of native species thinks so. He makes the case for an inclusive silvicultural approach. as we walk through a mixed-age and mixed-species forest, in nearby Laragh. "All the species here are native to our children," he says.

Fundamentally, Glendalough is managed as a multipurpose forest which provides wood and non-wood benefits. During my time here, it generated much-needed rural employment in a small community where most of the 80 or so forest workers and contractors in this tight community would have emigrated without the forest. The forest too has been acknowledged as a major recreation resource since the 1970s. This was the decade that an open forest policy was initiated and expanded over the years. Today, an estimated 38 million annual visits are made to almost 400 forests across the island. Like Glendalough, these demonstrate that while wood is the economic driver, other environmental and social benefits flow from sustainable forest management.

Since my time in Glendalough with the then Forest and Wildlife Service, the management of the oak woodland was handed over to the National Parks and Wildlife Service in 1989, with Coillte taking charge of the upland commercial forest. While the rationale behind this was understandable, it placed an unnecessary barrier between wood and non-wood aspects of forestry as if there was nothing in common between the native and the introduced; as if they need to be separated into conservation and commercial. This approach misses the point that both need to be managed sustainably to ensure their roles as renewable resources. Because we lost our native woodlands through reckless overcutting doesn't mean that they cannot produce wood sustainably for locally based furniture, turnery and other commercial enterprises. And because our commercial coniferous forests are primarily renewable wood producers, this doesn't preclude their non-wood functions.

Forest visitors don't draw a distinction between native and non-native – conifer and broadleaf. Nor should there be a division between their ecological function even though there are clear differences in emphasis. The lazy use of monoculture where it no longer exists and the even lazier use of invasive, where it never existed needs to be addressed in Irish forestry. This is not to dismiss claims of an overreliance on a single species when diversity is an obvious requirement, as our trees – our forests – are threatened with disease, insect damage and drought which are directly related to global warming. And when forests are threatened, so too are livelihoods. In this regard, Glendalough taught me one unassailable lesson: forests are first and foremost about people; the people who care for them, who depend on them for a living and the people who experience them as places of recreation and refuge.

Halfway into researching and writing this book, the country was in lockdown due to the Covid-19 pandemic. Those who could, worked from home, while essential workers maintained a semblance of normality during the worst global health and economic crisis in living memory. The role of the forest was crucial during this period. Record numbers of people visited our forests and woodlands to find respite; to communicate with nature during uncertain times. Most businesses closed down but forestry operated throughout Covid-19. Life went on for those who work the forests, as they were established, maintained, harvested and renewed, mostly out of sight in the eerie silence of the pandemic.

Forestry, like agriculture, was treated as an essential sector during the pandemic. Wood was needed for construction and a myriad of uses as the link between the forest and end product was maintained to produce even the most commonplace, but essential products such as renewable – and reusable – pallets. These were never more in demand for the storage and transportation of food, beverages, pharmaceuticals and other vital products.

The pandemic was an opportunity to re-assess the forest as a haven in times of stress and uncertainty. My own – and my family's – sylvan sanctuary during this period was the Devil's Glen Forest, or Glanmore as depicted in the sonnets of Seamus Heaney. Visits to this beautiful Coillte woodland when Covid-19 reached our shores in 2020 provided me with an opportunity as Rotary Ireland environment chair, to initiate a project which would renew and reimagine the benches along the forest walk to honour the Nobel Laureate. This project, initiated by a group of us foresters in the 1990s, was in need of renovation and despite the pandemic, Technological University Dublin (TU Dublin) students and lecturers undertook the task. They recreated the benches from locally sourced wood, safe in the knowledge that what they took was replaced in the forest. Then they etched Seamus's chosen lines of poetry, back in 2007, on the backrests before the benches were installed in 2022. That year had an added significance for me as 50 years previously I had made my own uncertain, but exciting journey, beginning in Glendalough Forest through the ancient native oaks and the giant Douglas firs. Like forestry back then, collaboration was essential in making it all work. This time round it required a partnership between the Rotary Club of Wicklow, Coillte, TU Dublin, the Department of Agriculture, Food and the Marine, local sawmiller Pat Staunton and the community, all with the support of the Heaney family. The homegrown Douglas fir wood – chosen for the benches – is the same species selected for transmission poles in Glendalough but this time round the emphasis is on the aesthetic rather than the industrial, although both qualities are required. Now it is chosen for its utility and beauty – the perfect material to rest on and take it all in, or as Seamus entreats us on one of the benches:

> *Walker, pause now and sit. Be quiet here.*
> *Inhale the breath of life in a breath of air.*

The pandemic, despite its upheaval, was an opportunity to make the connection between the living, breathing forest and what it says, what it produces and what it conserves, which is the essence of multipurpose forestry. I explore and seek answers to this connectedness, which is the purpose of this book. Despite advances in technology and science, we still need to return to nature, but with a renewed vision and an open mind.

In researching and writing this book, I owe a debt of gratitude to a number of people and organisations. Sincere thanks to the Forest Service, Department of Agriculture, Marine and Food, and the Society of Irish Foresters (SIF) for their support from the outset. I wish to acknowledge Gino Forte of Grasshopper Graphics for his perseverance and creativity in making sense of the ever changing tables, figures and text. Pat O'Sullivan, technical director, SIF was exceptionally generous with his time and objective criticism. He also painstakingly proofed a number of drafts while I also received advice and guidance from Gerry Murphy, ornithologist. My thanks also to Noel Gavigan, Donal Whelan, Neville Dukes and Des O'Toole. Thanks to all the staff at The Lilliput Press for their support and the various organisations I work with including SIF, *Irish Farmers Journal*, Irish Timber Growers Association, Rotary Ireland, stakeholders in Wood Awards Ireland and my fellow foresters I encountered along the way. Above all, sincere thanks to my family for their encouragement but especially to my wife Angela for her support, generosity of spirit and patience from start to protracted finish, not only for this project but many others over the years. Finally, I wish to acknowledge the people who nurture our forests, which are a welcome respite for all of us, especially when the body and soul need solace and renewal.

INTRODUCTION

Did you know our suburb was forest?
Our roof was a home for thrushes.
Our front door was a wild shadow of spruce
Eavan Boland, "VIII Once", *Code*, 2001

As we clear and burn forests, we release dangerous amounts of carbon back into the atmosphere, increasing the rate of climate change. We need to reverse this and create a world of expanding forests once more.
David Attenborough, *Our Planet: How to Save our Forests*, 2019.

The answers to "Why forests? Why wood?" have been interlinked since prehistory; to a time when the forest was valued for both its wood and non-wood benefits. As pressure increased to over-exploit forests during the eighteenth century in Europe and the twentieth century in the tropics, the resulting deforestation not only destroyed the forest as a renewable timber resource, but also led to soil erosion, flooding, destruction of habitat and loss of livelihoods.

Gradually Europe began to restore its forest resource so that today, forests cover is almost 40% of the land area. The destruction of Irish forests was more acute than in Europe as it continued unchecked over a much longer period. Today, forest cover is over 11.5% compared with 1% a century ago. If hedgerows, scrub and fragmented small wooded land are added, some 16% of the land area of Ireland is under trees.

The deforestation that Europe experienced was similar to the unsustainable – and often illegal – logging that has been taking place globally over the centuries and in tropical forests since the last century. Like the formerly overexploited European forests, communities in the tropics face prospects of wood shortages and destruction of habitat, but also additional catastrophes resulting from deforestation. These include desertification and global warming, which will have their greatest impact on communities living in the tropics, but the worst excesses of climate change will affect every community on our planet eventually, unless we reverse this trend as David Attenborough advocates.

The solution to reversing the destruction of these forest ecosystems lies in sustainable forest management. A number of countries that have experienced deforestation are slowly replacing and repairing the links in the forest and wood chain through sustainable management. This approach, which stresses renewability, acknowledges the importance of the relationship between each link in the forestry and forest products chain: from habitat protection to harvesting; from seed to sawlog; from tree replacement to tree felling. In this regard, the answers to "Why forests? Why wood?" ultimately lie in the maintenance of the forest chain, as the forest and its wood products are interdependent. So, the answers are found both inside and outside the forest ecosystem:

- In the forest, where carbon sequestration plays a key role in climate change mitigation; from tree canopy to herb layer where the visible meets the invisible soil under-world, which is the largest carbon store in the forest ecosystem.

- Outside the forest, where carbon storage in wood products displaces fossil-based materials with high embedded carbon such as steel, aluminium and concrete, as well as generating energy which displaces fossil fuels.
- End of life and beginning of new life, when wood can be recycled into new products or into carbon-neutral wood energy.
- Beyond wood, where new product development in the bio-economy creates further opportunities for wood in achieving a fossil-free society including energy, transport, medicine, textiles and other new areas of natural product development.

Sustainable forest management includes the protection of our natural forests and maximising the multipurpose objectives of mixed aged, mixed species forests. It also includes the creation of new commercial forests because "our growing global population will need to use more wood" as Sir David Attenborough explains. He has outlined the role of new productive forests and their relationship with the natural forests; where the need for sustainable wood is combined with economic, social and ecological objectives. Sir David Attenborough's planet allows natural and wood-producing forests to co-exist so that wood and non-wood multipurpose objectives can be achieved in harmony.

Wood benefits

The positive role of productive forestry in displacing fossil based materials in wood construction and energy is now being understood especially by innovative architects, engineers, scientists and wood processors who look to wood as a key resource in sustainable living.

The greatest benefits of forestry in twenty-first century sustainable living lie in their contribution to the bioeconomy – the area of the economy that uses renewable biological resources rather than fossil fuels. Pekka Leskinen, head of the Bioeconomy Programme, European Forest Institute (EFI) is convinced that "forests, forestry and the forest-based sector are the cornerstone of the European bioeconomy, and a major contributor to climate change mitigation".

Forestry has made a major contribution to the bioeconomy for centuries in construction, furniture, packaging, paper, renewable energy and other specialist uses. While these are likely to continue as the key markets, the role of wood in pushing the boundaries to achieve a green environment is being explored in new wood-based products. These include biodiesel and biogasoline, textiles, medicines and even food. Karl-Henrik Sundström, managing director, Stora Enso, believes decarbonising the economy will need focussed research in biorefining and biomaterial as well as construction of wooden buildings using engineered timber. Sundström, who is a member of Sweden's National Innovation Council, maintains that the transition to a fossil-free society is achievable as he believes that "everything made from fossil-based materials today can be made from a tree tomorrow".

Sundström's and Leskinen's confidence in wood as a key driver in sustainable living, stems from the performance of forestry in their own countries – Sweden and Finland – where a forest and wood culture is central to their nations' economic and social wellbeing. But it is often forgotten that even countries with the most assured wood cultures today, experienced deforestation as recently as the last century. For example, in the early 1900s, large tracts of Sweden's forests were cleared and degraded due to overcutting to satisfy the demands of industrialisation, increased population pressure, firewood demand and intensive farm cultivation and grazing.

A combination of State and private forest development has resulted in Sweden's forest cover, doubling in the past 100 years. Today, Sweden provides 10% of worldwide demand for sawn timber, pulp and paper products from forests that cover 1% of the global land area.

To achieve this performance, the Swedes harvest 90 million cubic metres (m^3) of timber each year from an annual forest increment of 120 million m^3. Sustainable forest management allows their foresters and forest owners to leave 30 million m^3 in the forest as a reserve to ensure the forest continuum is maintained and habitats are not overexploited.

Wood production is a key aspect of the European sustainable forest management model. The benefits of wood production, while acknowledged in Ireland, are still not fully understood, especially in its carbon substitution role. The carbon substitution or displacement factor describes how much greenhouse gas emissions can be avoided if wood and wood-based products are used to displace fossil-based materials. Substitution has only received a cursory mention so far in the climate change debate in Ireland, but it is a major benefit of trees in the bioeconomy especially in the use of engineered wood in large-scale construction as Colin Tudge writes in *The Secret Life of Trees,* where 'Timber cities would lock up a great deal of carbon'. But to achieve this "we need more wood" as David Attenborough explains, and by extension we need more forests. Forest expansion in countries that have lost their forest resource needs to be planned so it doesn't threaten other land uses, in particular sustainable agriculture. In Ireland, food and wood security are achievable as we have a sufficient land base to allow both to prosper.

Trees not only grow well in Irish climatic conditions, but a desired 60% increase in forest cover up to 2050 can be achieved without reducing agricultural production. This allows farmers to not only "farm trees" for much-needed wood, but also to protect and enhance our badly depleted native woodland resource. This approach allows forests to maximise their wood and non-wood benefits.

Non-wood benefits
The positive role of non-wood forestry is now acknowledged and understood. Sustainable management ensures that non-wood benefits flow, including biodiversity enhancement, recreation, landscape improvement, water protection and climate change mitigation.

The most widely appreciated non-wood function of the forest is its recreational role. The numbers of Irish people and tourists who visit our forests is increasing year by year. Although Ireland is one of the least forested countries in Europe, it has close to 400 recreational forests and woodlands open to the public. This is largely due to a visionary decision by the State to create an open forest policy in the 1970s. This allows the public to use forests for recreation and today this policy is endorsed by Coillte, the National Parks and Wildlife Service (NPWS) and some local authorities, while a similar approach is adopted by the Northern Ireland Forest Service (NIFS) and other agencies.

The importance of our forests as places of solitude was never more in evidence than during the outbreak of the Covid-19 pandemic in 2020 and 2021, not just in Ireland but throughout the world. Currently, 38 million visits are made annually to Irish forests by nature enthusiasts, seasoned and casual walkers who value trees as symbols of renewal across the island. When this resource was threatened in 2012 by government plans to sell the timber harvesting rights in Coillte forests, a series of nationwide peaceful protests was organised. Protesters voiced their fears that public access to Coillte's forests would be denied and the privatisation plan was wisely dropped in 2013.

The greatest benefit of forests lies in climate change mitigation by sequestering carbon dioxide (CO_2), which is increasing to unsustainable levels. Before considering questions on the relevance of forests, it is impossible to ignore the threat to the renewability of the global forest, which is central to climate change mitigation and sustainable living. "Forests are the lungs of the planet" sums up their importance in the climate change debate but it is in danger of becoming a hollow slogan while deforestation remains a reality. Despite our dependence on this great natural resource to live and to breathe, our forests are still being destroyed by illegal logging, deliberate burning and other indiscriminate clearances that contribute to global deforestation.

While sustainable management in Europe ensures the forest continuum, we have a responsibility to help countries struggling to protect their forests. In the past, Europe, including Ireland, played its part in global deforestation by importing wood and wood products without questioning their source, even though widespread illegal logging was taking place in the hardwood trade, especially in the tropics.

Against this backdrop, the questions posed in *Why Forests? Why Wood?* are as fundamental as ever. The dual role trees play in carbon sequestration within the forest and carbon displacement outside the forest is now widely acknowledged but has yet to feature in forest policy in many countries including Ireland. For example, the most obvious benefit of forestry in Ireland lies in its capacity to help farming achieve climate change mitigation targets as agriculture accounts for 33% of Ireland's greenhouse gas emissions. Switching as little as 2.5% of marginal agricultural land to forest every decade would achieve 17% forest cover in Ireland by 2050. This would ensure sustainable agriculture and forestry sectors.

Most of this afforestation programme will be carried out by farmers, and will provide wood and non-wood benefits. It will include diversity of tree species and diversity of objectives, to allow forest owners achieve a financial return from their productive coniferous forests while also enhancing and protecting the native and naturalised broadleaf resource.

Achieving the optimum balance between forestry and other land uses and the correct species balance in the protected and new forests are objectives facing not only forest owners but also society in general especially at a time of climate uncertainty. There is a shared responsibility in addressing the questions posed in *Why Forests? Why Wood?* This will require a partnership approach between the State and private stakeholder that acknowledges forestry as a multipurpose resource, whose value is measured not just in economic terms, but in its total contribution to society.

CHAPTER ONE

Forest, Woodland or Plantation? Defining Tree-Covered Land

Forest. *That's the word. Not a strange word at all but one he possibly never used. A formality about it that he would usually back away from.*
Alice Munro, "Wood", *Too Much Happiness*, 2010

Roy, the injured protagonist in Alice Munro's short story "Wood" is trying to remember another name for the familiar woodland that surrounds him after his tree-felling operation goes wrong. Waiting for help to arrive, he visualises it as "a tall word that seems ominous but indifferent". As Roy's wife Lea rescues him, he remembers: "*Forest*. That's the word." Roy concedes that it is not a "strange word at all," but compared with the words he uses such as "bush" and "wood", there is "a formality about it that he would usually back away from".

Another Munro character – Juliet – in the story "Chance" also struggles in naming "the mostly evergreens" she sees during a train journey from Ontario to Vancouver. She settles hesitatingly, but correctly, on the word *Taiga*, the boreal forest that stretches beyond Munro's native Canada to Russia and Scandinavia.

Munro captures the difference between Roy's intimate deciduous lowland woodland of cherry, ash, beech and oak, and the vastness of Juliet's coniferous "pine or spruce or cedar" in a mountainous landscape of "rocks, trees, water, snow". The difference is not only in tree species but also scale, topography and the soils that support the trees, and their underlying flora.

Which to use? Forest, wood, woodland or plantation? And does it matter? All are ecosystems dominated by trees but each has its own characteristics. Munro's forest is dark and remote "tall and ominous" which fits with Michael Allaby's definition as "a plant formation that is composed of trees, the crowns of which touch, so forming a continuous canopy". This contrast with his definition of a woodland as "a vegetation community that includes mature trees which are usually spaced more widely and so are more spreading in form than forest trees [i.e. their crowns are not touching and they do not form a closed canopy] ... often defined as having 40% canopy closure or less". On the other hand, Oliver Rackham states: "Woodlands [comprise] trees so close together that their canopies meet...trees

IRELAND AND EU FOREST DEFINITIONS

The legal definition of a forest in Ireland appears in the Irish Forestry Act 2014 as: "Land under trees with: (a) a minimum area of 0.1 hectare (ha), and (b) tree crown cover of more than 20% of the total area, or the potential to achieve this cover at maturity, and includes all species of trees". In the EU, "forests are deemed to be land with an area of more than 0.5ha and tree crown cover of more than 10%, and where trees can reach a minimum height of five metres at maturity". Both the Irish and EU definitions allow for areas so small that they could best be termed groves or copses. Both definitions allow for agroforestry where tree cover is as low as 400 trees per ha and where tree growing is combined with grazing, silage, fruit and other crops.

are managed by coppicing or allowed to grow on into timber."

It follows that a wood or woodland is a tree-covered area, generally more open than a forest and comprising mainly broadleaf tree species compared with the more closed *selva oscura* which Dante enters as he opens the *Inferno*:

> *Nel mezzo del cammin di nostra vita*
> *mi ritrovai per una selva oscura*
> *che la diritta via era smarrita*

Yet, in all but nine of the 50 English translations – so far – *selva* (from the Latin *silva*) is translated as wood rather than forest to describe Dante's dark, foreboding surroundings. Henry Wadsworth Longfellow is one of the minority to opt for forest:

> *Midway upon the journey of our life*
> *I found myself within a forest dark,*
> *For the straightforward pathway had been lost.*

That Longfellow chooses forest rather than wood is not surprising. The first American to translate *Divine Comedy*, he is more in tune with the deep, dark coniferous forest, where one actually risks getting lost. His forest inscape described in *Evangeline: A Tale of Arcadia,* has similarities with Dante's *Inferno* in its dark and ominous opening:

> *This is the forest primeval. The murmuring pines and the hemlocks,*
> *Bearded with moss, and in garments green, indistinct in the twilight,*
> *Stand like Druids of eld, with voices sad and prophetic,*
> *Stand like harpers hoar, with beards that rest on their bosoms.*
> *Loud from its rocky caverns, the deep-voiced neighbouring ocean*
> *Speaks, and in accents disconsolate answers the wail of the forest.*

In Ireland, translating wood, woodland and forest is flexible; the words are often interchangeable. Irish translations include *foraois* (forest of timber trees), *coille ard* (open or high forest) and *coillearnach* (woodland). The Irish word *coill* can mean either forest or wood. "*An choill a thabairt ort féin*" literally means "to take to the woods" but also "to become an outlaw" where the wood is a hiding place. It becomes a place of refuge for the wood kernes, medieval Irish foot soldiers, described by Cardinal Beufort in Shakespeare's *Henry VI* as "The uncivil kerns of Ireland".

The wood as a place of refuge is referenced as far back as the seventh century when Sweeney, the northern king, fled to the woods after the Battle of Magh Rath in 637BC. Seamus Heaney drew political, poetic and societal parallels with Sweeney's travails, to map his own journey from Northern Ireland to Glanmore (near the Devil's Glen Forest), outside Ashford, Co. Wicklow as "a wood-kerne / Escaped from the massacre".

Elsewhere, the definitions of woodlands and forests are found in their language source. Woodland, originates from *wald*

Gustav Doré's engraving of Dante's Divine Comedy *depicts the forest as a dark foreboding inscape.*

or *weald*, both Anglo-Saxon words for "open", while forest is derived from the Latin *foris,* meaning "out of doors". This refers to unenclosed land or the land that lies beyond fenced or enclosed cultivated land. In this context the meaning of forest or its derivation *foris* is obvious. Forests were associated with poorer land and generally comprised conifers on the exposed uplands as opposed to woods or woodlands, which mainly carried broadleaves in the lowlands and often lay between farming land and the forest.

Conifers grow on the poorest sites and the highest altitudes, only ceasing to grow above the treeline where soils are too shallow and temperatures too low to support forests. The definition of forest as land on the outside, or outside managed farm land, is apparent in North America and most European landscapes. The distinction between forest and woodland is less clear in Ireland, mainly because the countryside was almost denuded of forests and woodlands from the seventeenth century and probably much earlier.

The large tracts of pine forests that covered much of the landscape began to die off with the spread of bogland at least 7,000 years ago and were virtually wiped out by the early Christian period. The pine forests that grew at high altitudes are long gone, so in Ireland we have little concept of the natural forest or woodland. The little we have are semi-natural indigenous woodlands, vital in providing an understanding of our woodland heritage, but requiring conservation and reimagining to make the connection over time with their impressive primeval ancestors. Native woodlands now cover 182,000ha, but these comprise semi-natural woodlands and mixed woodlands along with very young plantations established from as recently as the turn of this century, when a quarter of the native woodland resource in Ireland was planted.

The introduction of plantation to the lexicon describes young, productive tree-covered areas. In Ireland, plantation has negative historic connotations as it is associated with the sixteenth- and seventeenth-century confiscation of Irish land by the English crown the colonisation of settlers ("planters") from Scotland and England. Plantation also has negative silvicultural nuances, especially when the term is combined with monoculture(s) to describe contemporary forestry practice. While monocultures featured in Irish forestry during the period 1950 to the 1990s when two conifers – Sitka spruce and lodgepole pine – dominated State afforestation on poor land, this changed from the early 1990s when better quality land was made available for afforestation, mainly by farmers. This resulted in a change in species composition from 95% to as low as 62% conifers during afforestation programmes from 2000 to 2012. Now, new forests must include 35% broadleaves and open spaces comprising roads and trails, firelines, rivers, streams and aquatic buffer zones with occasional small trees and shrubs and other flora. The average size of afforested land today is less than 8ha and comprises at least two species and open areas so the term monoculture is no longer relevant. Yet, the term plantation has some currency when applied to Irish forestry as almost half the stocked forest estate at the time of writing is less than 20 years old. Afforestation of 105,000ha of conifers during the first two decades of this century, is likely to have rotations of little more than 30 years but 45,000ha of mainly native broadleaves were also planted some with rotations that will extend to the next century. In addition, some of the coniferous plantations may be converted to continuous cover forest management (CCF), so regardless of species, at some stage a plantation can make the transition to woodland or forest depending on the tree species and the management objectives.

Regardless of the definition, while trees are the key components, they are only part of the forest ecosystem which includes other vascular plants and flora, fauna and microorganisms, both native and introduced. Humans are also an integral part of the forest ecosystem. Their economic, environmental and social needs influence the sustainability of all forest ecosystems. Their requirements can either enhance or damage the forest ecosystem, but the view that natural woodlands should be left untouched to guarantee their biodiversity and longevity is misplaced – certainly in temperate forests, less so in tropical forests. Even after exploitation, temperate forests can be restored as the soil will withstand a high degree of disturbance, while the nutrient-poor soil in a tropical forest, mainly comprising rapidly decaying organic matter, is often unable to recover after large-scale clearcutting and fires.

Placing humans in the forest ecosystem is an acknowledgment that all forests need management. Even, the Białowieza Forest, the last remaining large tract of primary forest in Europe, which straddles Poland and Belarus, has to be

The primeval forest of Ravna vala, southeast of Sarajevo in Bosnia and Herzegovina, where silver fir and beech form the two main species over the millennia. Trial research plots in this forest demonstrate that there is greater understorey and tree species diversity in stands where continuous cover forestry (CCF) is practised than in stands that are left to natural undisturbed regeneration.

SEMI-NATURAL WOODLAND

Woodland is the best term to describe the Vale of Clara Nature Reserve, Co. Wicklow. It's tempting to call it a natural woodland that has a direct link with the primeval. Unfortunately that connection has been broken not only in Clara, but throughout Ireland and virtually all of Europe. It is a semi-natural woodland, as the oak was once cleared and the Scots pine was introduced from Scottish stock. Clara is close to the primeval not only in the canopy trees but also in the secondary species such as birch and mountain ash, and the herb layer. But Clara also has invasive species including western hemlock (above) and a beech understorey (overleaf).

Those who advocate leaving the wood to nature, forget that without management, these species will eventually outcompete the oak and Scots pine while the herb layers will be over-browsed by Sika deer. Without management, the Vale of Clara would morph from an indigenous woodland to a non-native forest. Conservation is vital as Ireland's moist mild climate is conducive to all tree and shrub growth, as witnessed in Killarney where *Rhododendron ponticum* threatens the oak woodlands.

Overleaf: Oak woodland, Vale of Clara, with beech understorey. Unless this woodland is managed, the non-native beech will become the dominant species as the wood transitions from native to naturalised forest.

managed and monitored to ensure the equilibrium between forest, wetlands, agriculture and river corridors is maintained. Without a sustainable conservation management plan, the region's flora and fauna would be threatened not only from within but also from outside, so a natural buffer zone is maintained to protect the Białowieza from external threats.

Likewise, the virgin forest of Ravna vala, southwest of Sarajevo in Bosnia and Herzegovina requires management. This haven of biodiversity has two main tree species – silver fir and beech – with a range of other species including Norway spruce, sycamore and maple vying to be part of the post-primeval narrative. Beech and silver fir are shade bearers, so they form the main crop in the virgin forest as they maintain an equilibrium which has lasted down through the millennia. In areas where there is no management, the trees are allowed to die and rot, renewing the soil and the rich flora. The surrounding forest is managed according to the principles of CCF, which creates openings in the canopy and allows greater scope for Norway spruce, sycamore and maple to develop as these are light demanders. In this instance, the ecosystem benefits from human intervention, resulting in greater flora biodiversity than the untouched primeval forest.

Unlike the Białowieza and Ravna vala forests, Ireland has lost all contact with the primeval forest, but our fragmented semi-natural woodlands provide sufficient information to re-imagine and recreate our lost forest resource. But to do so, they too require intensive management if they are to survive as unique ecosystems because all are vulnerable to invasive tree and shrub species as well as mammals that have been introduced over the centuries.

For example, the native oak in Killarney, if left untouched, will be damaged by browsing Japanese sika deer and colonised by invasive rhododendron (*Rhododendron ponticum*) which will destroy the native woodland habitat. Likewise, the oak and Scots pine in the Vale of Clara will not be able to compete with exotic conifers such as western hemlock and naturalised but non-native beech unless there is human intervention. European beech and western hemlock from western North America are shade bearers so they can survive under the canopy of native trees, before eventually making their way upwards, outcompeting the native oak and Scots pine trees as well as killing off the shrub and herb layers on the forest floor. Without controlling these two species, the Vale of Clara woodland would eventually make the transition to a predominantly beech and hemlock forest with the final battle for supremacy taking place between these two European and American outliers. But is it fair to call these two beautiful trees outliers? Beech has been in Ireland possibly for seven centuries while western hemlock is a more recent arrival, having been introduced in 1854. Beech is non native but naturalised, while western hemlock is an exotic but robust natural regenerator, so both have developed a staying power in what was once, but is no longer, an alien landscape.

Farm or forest? Agroforestry is rec-ognised as an alternative to full forest cover as it allows farming and forestry to coexist. It conforms with the definition of a forest as it carries a minimum tree density of 400 trees per ha.

The transformation from plantation to multipurpose forest is a relatively fast process in Ireland as illustrated in Coillte forests in Co. Wicklow. The Devil's Glen (opposite) combines recreation with production forestry by opting for a thin and clearfell re-gime on exposed sites with CCF prac-tised in sheltered areas. The second rotation Douglas fir and Japanese larch crop is over 30 years old while the CCF area is over 70 years old. The commercial forest (below and page 4), situated across the Avon-more River from the Vale of Clara Nature Reserve, was planted mainly with Norway spruce and Scots pine. Over the years it has transitioned from a plantation to forest, while open areas along the river bank are colonised with naturally regenerating willow, birch and holly.

It is essential that the native Vale of Clara is protected, but its conservation brings into sharp relief how we treat our introduced species. Where do they belong in a landscape that is largely artificial, including our field system, cleared land for agriculture, and our introduced agricultural crops and live-stock?

A new forest type is now entering this landscape. Agrofor-estry, the combined use of agriculture and forestry on the same plot of land, is rare in Ireland but not unusual in continental Europe and elsewhere (see also Chapters 5 and 9). In tropical countries it has wider aims including soil stabilisation, shelter and soil replenishment. In suitable holdings, agroforestry is rec-ognised as an alternative to full forest cover. It allows main-stream farming to take place alongside forestry, and has the potential to produce high-value timber crops.

The definition of wood, woodland or forest in this new order is open to interpretation and the sense of place these tree-covered places evoke. Traditional naming of places may decide, but ultimately visitors surrounded by trees will know from the visual and spiritual experience what it is they are standing in. If it has open or low-density mainly broadleaf trees with good ground cover of vegetation, then it's a woodland. If it has the formality described by Alice Munro, with more closely spaced coniferous trees and less ground cover, then it is a forest.

"Save our forests" was the successful rallying cry of those who opposed the Irish Government's planned sale of Coillte's harvesting rights in 2011. The public identified with the state owned company's recreation forests and "open forest" policy. In this context forest is perhaps the correct word as it em-braces the native, naturalised and exotic just as it acknowl-edges our multicultural society.

CHAPTER TWO

Forests Past, Forests Present: The Uncertain Narrative of Irish Forestry

Our land, our shelter, our woods and our level ways
are pawned for a penny by a crew from the land of Dover.
Aogán Ó Rathaille, "No Help I'll Call", *Cabhair ní Ghairfead,*
c1729, translated by Thomas Kinsella, 1981

With the disappearance of the woods, the sawmill in many a locality
vanished too. With it went the custom of using home-grown timber and
the lore connected with it.
H.M. Fitzpatrick, *The Forests of Ireland,* 1965

The softening ruts

lead back to no
'oak groves', no
cutters of mistletoe
in the green clearings.
Seamus Heaney, "Bog Oak", *Wintering Out,* 1972

Sweeney at Drimcong, Brian Bourke's depiction of Suibhne mac Colmáin, the Ulster King of Dal Araidhe who is cursed to wander as a birdlike creature "among dark trees/between the flood and ebb-tide/going cold and naked" according to Seamus Heaney's translation Sweeney Astray *which he believes was "taking shape in the ninth century". It includes a hymn-like roll call of trees as Sweeney "praised aloud all the trees of Ireland".*

To gain an understanding of Irish forestry today, it is important to reconnect with forests past. The destruction of Irish forests until the beginning of the last century, resembles more recent deforestation in the tropical countries, albeit over a longer timeframe. Nevertheless, the final and brutal act in Ireland's forestry narrative is unique in Europe. Yet, despite H.M Fitzpatrick's bleak assessment of early twentieth century Irish forestry, there is sufficient evidence available to reimagine a future forestry paradigm and, judging from some contemporary innovative wood architecture and design, we have not altogether lost "the lore" of using wood. Fitzpatrick was referring to the widespread overcutting of the few remaining forests when forest cover eventually fell to 1% of the land area of Ireland. Now, a century later, forest cover has increased tenfold. Forest cover today – at 11.5% – is more than four times greater than it was, even as far back as the mid-seventeenth century.

So, despite the destruction of Irish forests over the millennia, lost ground has been recovered, but the longing persists: to regain something akin to the ancient forests. Although primeval forests have been lost in most European countries, the yearning for renewal is greater in Ireland. This is partly due to the scale of the loss, which can only begin to be assuaged by first exploring the clues available, to reconnect with the primeval forest. While famine, poverty, war and loss – of both land and language – are commemorated widely in Ireland, loss of forests and their accompanying culture have received little attention. Efforts to simplify – and explain – this loss by delving no deeper than the opening lines of the nineteenth-century poem "Cill Chais", although irresistible, have proved woefully inadequate. The demise is far more complex, as the evidence of a once abundant wooded landscape has been etched from memory for centuries.

PRIMEVAL

The physical evidence of forests past, is found in the few remaining semi-natural woodlands, such as the Vale of Clara and Glendalough, Co. Wicklow; Glengarriff, Co. Cork; St. John's Wood, Co. Roscommon, Killarney, Tomies and Uragh woods, Co. Kerry; Brackloon, Co. Mayo; Abbeyleix, Co. Laois and a few fragmented native woods. But even in these, the link with the primeval is tenuous. For example, the native woodlands that cloak the lower slopes at Glendalough today didn't exist 150 years ago apart from a few isolated trees. The remaining evidence of the primeval forest is buried in bogs. In this underworld of Seamus Heaney's "Atlantic seepage", Ireland's woodland prehistory is concealed and revealed in preserved tree stumps or in pollen grains conserved in layers of acidic peat. In "Bogland", Heaney poetically explores and excavates this underworld, where he finds:

> *Only the waterlogged trunks*
> *Of great firs, soft as pulp.*

   These provide us with valuable information about the primeval forest, its scale and its people who arrived here much earlier than previously thought. Until relatively recently, it was believed that the earliest human settlements in Ireland date to no earlier than 8000BC. In 2016, radiocarbon dating of a brown bear bone with knife marks – found in a Co. Clare cave in 1903 – confirmed that there were humans in Ireland during the Palaeolithic or Stone Age era. The existence of humans in Ireland some 12,500 years ago is 2,500 years earlier than previously believed. How these and later visitors during the Mesolithic (8000-4000BC) and Neolithic (4000-2500BC) periods interacted with the forests that covered most of the island is open to conjecture. We know from variations in pollen grains of different trees species that the earliest Neolithic farmers slowly changed the landscape from forest to agriculture. Was this clearance indiscriminate or was there a degree of order and respect for the then formidable trees oak, ash and pine forests? The giant primeval oaks would have provided continuity,

The Lurgan Longboat or Canoe. Dating to around 2500BC, this Bronze-Age boat was discovered in Lurgan Bog, Co. Galway in 1901. Its function is debatable. It may have been used for fishing or trading by river or lake. Probably unsuitable for sea travel, it may even have been built as a ceremonial artefact with aesthetic rather than functional objectives, although this is unlikely.

longevity and a sense of place in the consciousness of those who worshipped them or took shelter beneath their huge canopies. The social and spiritual importance of these revered trees has been celebrated in folklore, antiquity, religion, and literature – especially poetry – to such a degree that begs the question: "Why then did the forests virtually disappear?"

While trees were revered, they were also burned and cleared by early farmers to make way for agricultural land, not unlike present-day deforestation of the Amazon region. Although trees were valued for their spiritual symbolism, they were also prized for their material uses including heating, construction, boat building, bridge building and conversion into farm implements and weaponry. They also had non-wood benefits including food, medicine and shelter. So, while individual oaks may have been worshipped, oak woodlands were also felled, hacked and shaped to meet the needs of the worshipper.

The hollowed-out canoe in the National Museum illustrates the enormous bulk of specimen oak trees. Measuring 15m long and 1.12m wide, it dates to 2500BC. It is likely that a boat of this size would have been fashioned from a tree with a straight, largely knot-free bole, up to 20m in length. Allowing for crown spread, the tree that was selected to make this canoe may have been close to 50m tall. Today only two oak trees in Ireland have achieved heights of 37m. We can but speculate on the search to locate this mighty oak tree deep in the primeval forest and how long it took to fell it with crude axes. We can only imagine how the woodsmen felt when it crashed to the ground before setting to work to make the canoe, with plenty of timber left over for implements, furniture and firewood to cook and keep them warm over the winter months. We know that this event took place some 4,500 years ago during the formation of the Lurgan Bog in Co. Galway, where the boat lay buried until Patrick Coen discovered it in 1901 before it was transported to the National Museum.

This is the primeval world explored by Seamus Heaney in "Bog Oak" where the recycled bog oak remerges millennia later, this time as, "A carter's trophy / split for rafters". In this resurrection of the primeval we gain some understanding of the loss of a woodland culture, recreated poetically by Heaney. But, in the millennia leading up to the Christian era, we have no understanding of how the decline and fall of Ireland's once great forests was viewed by an increasingly agrarian people, as the artefacts and cultural memory have almost been wiped out.

But the memory of trees and forests endures in folk memory and language. The tree-influenced archaic script that appears on Ogham stones may precede Christianity, although scholars are divided on its origin. James Carney maintains that Ogham predates the arrival of Christianity by more than 500 years, while Damian McManus claims that it is a Christian script. However, it is not inconceivable that a literary and oral culture would have taken root in some form prior to the Christian era. By the early Christian period, humans had

Felling a mighty oak to make the 15-m Lurgan Canoe was a huge undertaking for a Bronze-Age people, because of its scale, and its status as a revered tree. (The white cutout shape indicates the dimensions of the canoe in proportion to the tree.) It would have required careful planning and intense labour in felling, followed by chiseling and sculpting to shape it into a functional vessel.

inhabited the island for 11,000 years, so communications must have evolved in language and inscription. Regardless of the Ogham script's origin, all agree that it features trees, but the popular belief that it is a "tree alphabet" with most letters having a corresponding tree name has been discounted. It is now accepted that at least eight are represented, namely: B (*beith* or birch); C (*coll* or hazel); D (*dair* or oak); F (*fearn* or alder); I (*idar, ídad, iodha, idhadh* or yew); O (*onn* or ash); S (*sail* or willow); and H (*huath* or hawthorn). While the question of origin is still an open one, Ogham is the first evidence of script and language that we have. Frank Mitchell and Michael Ryan are in no doubt that:

> The oghams, the first inscribed monuments in Ireland, bring us to the threshold of history and to the changes ushered-in by the arrival of Christianity.

Although forests were in decline for most of the first millennia, trees continued to have a major influence on the lives of people in the seventh and eighth centuries as demonstrated in many placenames of the period featuring trees. This does not necessarily prove that Ireland was still heavily wooded during this period as isolated copses and prominent individual trees would no doubt have influenced placenames, as would the network of hedgerows that forms field and townland boundaries.

The tree retains its important presence in the development of an early legal scheme which tells us how society was structured under Brehon Law, administered by brehons, who were the successors to Celtic druids. Trees, featured in the Old Irish Tree List, are ranked not just by timber utility, but also by non-wood, medicinal, food and spiritual benefits. In the section of *Bretha Comaithchesa* or "Laws of the Neighbourhood", dealing with trees and woodlands, the Old Irish Tree List acknowledges a hierarchical grading system for trees. For example, oak is a tree of great prominence and regarded as a 'noble of the wood' because of its wood and non-wood properties, as is pine (now Scots pine). Yew was also rated Class I, not only because of its beautiful wood but also because of its spiritual associations with life and death and possible medicinal properties. Hazel and apple were regarded as Class I trees because of their importance as food producers ahead of more important timber trees such as elm and alder.

The forest (*coil, foraois*) as represented in Irish language derivatives of wood and woodland, reflects extensive usage as demonstrated by Gregory Toner and fellow authors. These include *feadhan* (spear-shafts), *chúach* (drinking vessel), *ciuil* (wooden instrument) and *crannach* (items made of wood, including logs for road paving).

While the physical link with the primeval forest is lost, there is an abundance of references in folklore and literature. Despite long silences and gaps in our knowledge, there are sufficient sources that link the poetic to the arboreal when the

THE OLD IRISH TREE LIST
From *Bretha Comaithchesa* or Laws of the Neighbourhood

CLASS I
Airig fedo – nobles of the wood
Dair – oak
Coll – hazel
Cuilenn – holly
Idar – yew
Uinnius – ash
Ochtach – pine (now Scots pine)
Aball – wild apple

CLASS II
Aithig fedo – commoners of the wood
Fearn – alder
Sail – willow or sally
Scé – whitethorn or hawthorn
Cáerthann – rowan or mountain ash
Beith – downy and silver birch
Lem – elm
Idath – wild cherry*

CLASS III
Fodla fedo – lower divisions of the wood
Draigen – blackthorn
Trom – elder
Féoras – spindle tree
Findcholl – whitebeam
Caithne – arbutus
Crithach – aspen
Crann fir – juniper*

CLASS IV
Losa fedo – bushes, plants of the wood
Raith – bracken
Rait – bog myrtle
Aitenn – furze, gorse or whin
Dris – bramble
Fróech – heather
Gilcach – broom
Spín – wild rose* or thorn tree

Source: Adapted by D. Magner (2016) from Fergus Kelly (1997) Early Irish Farming.

*Translations uncertain.

THE IRISH TREE ALPHABET

Note: This illustration of the Oghan alphabet is based on a paper "On the Ogham Character and Alphabet" by Charles Graves (1812-1907) published by the Royal Irish Academy (1847-1850). Graves also depicted aspen (ea), spindle (oi), woodbine (ui), gooseberry (ia) and hazel (ae). It is now accepted that the eight letters B, C, D, F, I, O, S and H owe their origins to tree or shrub names but the remaining letters may represents other objects or materials or may have meanings that have fallen out of use.

I	I	Yew	*Idhadh*
E	E	Aspen	*Eadhadh*
U	U	Heather	*Ur*
O	O	Gorse	*Onn*
A	A	Scots pine	*Ailm*
R	R	Elder	*Ruis*
Z	Z	Blackthorn	*Straif*
Ng	Ng	Reed	*Ngedul*
G	G	Ivy	*Gort*
M	M	Vine	*Muin*
Q	Q	Apple	*Queirt*
C	C	Hazel	*Coll*
T	T	Holly (?)	*Tinne*
D	D	Oak	*Dair*
H	H	Hawthorn	*Huath*
N	N	Ash	*Nín*
S	S	Willow	*Sail*
F	F	Alder	*Fearn*
L	L	Rowan	*Luis*
B	B	Birch	*Beith*

poet is in tune with a wooded landscape. These demonstrate that the destinies of trees and poetry are interlinked, as the poet leans on the tree as a reference point but there are periods when this connectedness is absent.

EARLY CHRISTIAN

Early Christian poetry is a rich source of information on trees and woodlands. While most of these lyrics weren't written down until the twelfth century and later, some can be traced to at least the fifth century. They illustrate a rich forest culture, but by then, much of the oak and almost all of the pine forests had been cleared or submerged in bogs.

Trees are central to much of early Irish verse which has a strong empathy with the natural environment. Of the 70 poems in Maurice Riordan's *An Anthology of Early Irish Lyrics*, 31 feature trees and woodlands. The poet refers to the surrounding woodland as a pleasant place to work and a source of inspiration. The tree is acknowledged for its spiritual and aesthetic significance, but also for its utility; as a source of wood and non-wood produce including medicine, food and drink. In Paul Muldoon's translation of "Marbhan and Guaire", the poet acknowledges hawthorn's medicinal qualities as being "good for a pain in the heart" and for refreshment he delights "to quaff a cup of hazel-mead". The anonymous poet who records the travails of the banished Sweeney acknowledges trees for their non-wood benefits such as shelter, as well as their functionality for producing weaponry:

> *Holly rears its windbreak,*
> *a door in winter's face;*
> *life-blood on a spear-shaft*
> *darkens the grain of ash.*

Those lines from *Sweeney Astray* by Seamus Heaney, are based on *Buile Suibhne*, "The Madness of Sweeney", by James George O'Keeffe. The basis of O'Keeffe's version is a seventeenth century manuscript, composed between 1200 and 1500, but Heaney believes the epic poem "was already taking shape in the ninth century".

In a number of early Irish lyrics, the poet is positioned in a benign wood such as the anonymous monastic poet who penned "The Scribe in the Woods". It was written down sometime between 800 and 850 on the margin of Priscian's *Latin Grammar*. Translated by Gerard Murphy, we cannot be sure if it was intended to be read as a poem. It has however been translated as a poem by Seamus Heaney, Ciaran Carson, James Carney and Máire MacNeill. The opening lines suggest an idyllic working environment as captured by Carney:

> *A wall of forest looms above*
> *and sweetly the blackbird sings;*
> *all the birds make melody*
> *over me and my books and things.*

Page 203 from St. Gall Priscian Glosses Latin Grammar *ninth-century manuscript. Note the insertion of "fidbaide" in the bottom margin (detail, page 36), translated as "hedge of trees" by Gerard Murphy; "green branches", by Maire MacNeill; and "greenwood trees" by Ciaran Carson in their translations.*

MacNeill continues the woodland and birdsong theme in a convincing second stanza:

> *The cuckoo pipes a clear call*
> *Its dun cloak hid in deep dell:*
> *Praise to God for this goodness*
> *That in woodland I write well.*

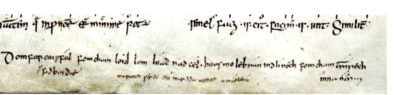

"Dom-fharcai fidbaide fál"

Dom-fharcai fidbaide fál
fom-chain loíd luin, lúad nád cél;
h-úas mo lebrán, ind línech,
fom-chain trírech inna n-én.

Fomm-chain coí menn, medair mass,
hi m-brot glass de dingnaib doss.
Debrath! nom-Choimmdiu-coíma:
caín-scríbaimm fo roída ross.

Anon. eighth to twelfth century. From *Early Irish Lyrics*, Gerard Murphy (ed), 1998.

This benevolent depiction of woodland, allowing the poet to work outdoors, may be attributable to climate change when "the golden age of the monasteries coincided with the gradual warming that led to the Medieval Warm Period from around 950 to 1250", according to Maurice Riordan.

Trees in Ireland reveal a strong sense of locus, especially in the naming of places since the pre-Christian era. Most place-names were probably fixed in the eighth century although the practice "was in active operation" up until the fifteenth century, originating from "names derived from English families" according to P.W. Joyce. Trees, forests and woodlands feature in an estimated 13,000 placenames in Ireland. This illustrates a strong affinity with trees and a robust forest and wood culture, but as stated earlier, may not necessarily be an indicator of widespread tree cover in Ireland at the time.

For all these references, there is little tangible evidence to confirm that Ireland was heavily wooded during the early Christian period. Valerie Hall maintains that the decline of Irish forests was well underway by the end of the first millennium. In a catch-all generic summary of forest decline, she maintains the land "was gradually cleared of trees over the last one thousand years, first by the Gael and then by the Planter".

That there was a gradual decline beginning with the Gael is true, but later political and industrial upheaval increased deforestation. For example, unsustainable demand for wood to satisfy shipbuilding and production of wine cask staves accelerated in the sixteenth century. Oak forests came under pressure to supply charcoal for industrial iron smelting, while ash proved to be an ideal species to fuel furnaces for glass production. Without a reforestation strategy, woodland deterioration and deforestation intensified.

POST KINSALE – TREES OF THE DISPOSSESSED

If all was sweetness and light during early Christian tree and woodland poetry, there was a dramatic change in the poet's perception of the dwindling forests especially after the Battle of Kinsale in 1601, when the Gael was swept aside. The defeat of the Irish and Spanish forces, and its aftermath, marked a seminal period in Irish history that reverberates politically, religiously and linguistically to the present day. There is no reason to believe that Ireland's forests escaped the fallout from the defeat at Kinsale. The Flight of the Earls and the end of the Gaelic Order were accompanied by the demise of the Bardic Order as depicted by Eavan Boland four centuries later:

After the wolves and before the elms
The Bardic Order ended in Ireland

Only a few remained to continue
A dead art in a dying land ...

The chronology may be confusing in Boland's poem as wolves didn't become extinct until the late eighteenth century, while most of the elm had died out by 3000BC. The apocalyptical note struck by Boland – "A dead art in a dying land" – indicates a death notice or at least an ending of sorts. Boland follows in the tradition of a long line of poets in using the tree as a symbol of loss and renewal. Elm is the perfect species that both defines and defies an ending. The species declined in numbers long before Kinsale due to over-exploitation and disease, possibly Dutch elm disease. Boland's "before the elms" may be referring to its reintroduction in the eighteenth century, when it continued to thrive until the second half of the last century before disease struck again and virtually wiped it out.

The acceleration of woodland exploitation took place during the seventeenth century when the first concerns for security of wood supply were raised. What follows Kinsale is the further decline of the forest which coincides with a decrease in poems that previously celebrated trees and woodlands. And when forests are featured, they are invariably linked to the decline of the Gaelic order, and loss of language, land and patronage for an embittered bard. The forest is now a hiding place for the landless wood-kerne (*ceithearn*) or tory (from the Irish *tóir* or *tóraidh* – not to be confused with later political usage). This is the battleground where hostilities are acted out between the dispossessor and the dispossessed – and within the dispossessed – as portrayed in a nursery rhyme dating from the period after the failed Irish Rebellion of 1641 and the Cromwellian Conquest of Ireland 1649-51:

Ho brother Teig, what is your story?
I went to a wood and shot a tory;
I went to the wood and shot another;
Was it the same, or was it his brother?

I hunted him in, and I hunted him out,
Three times through the bog out and about,
Till out of a bush I spied his head,
So I levelled the gun and shot him dead.

A disappointing aspect of the decline of a Gaelic culture – woodland, linguistic and poetic – is how much of it is missed by poets and commentators of the time. The popular poem that references woodland destruction is "Cill Chais". The poem is in the Bardic tradition and uses the felled forest as a metaphor for the demise of Gaelic Ireland and the Butler family in particular even though the actual chronology is askew. Placing "Cill Chais" in the Bardic milieu fits the poetic narrative but not the reality as "the last of the woods" (*deireadh na gcoillte*) in the poem were sold between 1797 and 1801 while the earliest manuscript appears a half century later. As "Cill Chais" is a retrospective lament of the collapse of Irish woodlands and the Gaelic order, we have to look elsewhere to find a more authentic voice to depict actual woodland decline.

"Thus go the Soldiers in Ireland, Beyond England, Thus go Peasants in Ireland", Albrecht Dürer, 1521. Woodcut, National Museum, Berlin.

The well equipped soldiers may be the gallowglasses or mercenary Irish warriors fighting abroad while the roughly dressed and poorly equipped peasants may represent the wood kernes who emerged from woods and forests to ambush the enemy.

Aogán O Rathaille (c.1670-1726) lived through the decline of the Irish language and the Bardic tradition. Unlike the anonymous composer of "Cill Chais" he is also an actual witness of woodland destruction. (For a time, it was believed that Ó Rathaille composed "Cill Chais".) Ó Rathaille's trees find expression in the *aisling*, or dream poem. "Dream" – translated by Michael Hartnet – is close to early Christian nature poetry in its benign depiction of trees and landscape:

> *A magical haze they'd arranged where no darkness appeared,*
> *from Galway of bright-coloured stones to Cork of the quays,*
> *nut-clusters and fruit grew forever on trees,*
> *acorns eternal in woods and honey on stones could be seen.*

The *Aisling* allows coded expressions of Ireland (usually as a beautiful woman). This land of bountiful trees and seasons is replaced with a harsher landscape in his uncoded political poems. These, not surprisingly, are bitter, not least because his once high status as a Bardic poet is now a distant memory. But he *is* a witness to forest decline:

> *A sharp grief to me the woundings of Ireland*
> *oppressed under clouds and her people all heart-sick;*
> *the trees that were strongest at giving them shelter,*
> *their branches are lopped, their roots withered and rotten.*

Ó Rathaille's resentment is tangible, his sense of loss agonising, as he is forced to live out a life in a cursed post Cromwellian landscape as outlined in "The Ruin that Befell the Great Families of Ireland":

> *Land that produces nothing of sweetness,*
> *land so sunless, so starless and so streamless;*
> *land stripped naked, left leafless and treeless,*
> *land stripped naked by the English bleaters.*

Ó Rathaille's poem is a political and environmental lament for a forest and its habitat in the face of political upheaval, and possibly climatic stress. Ireland experienced "The Little Ice Age" when the temperature of the Northern Hemisphere – from

Ó RATHAILLE: WITNESS OF FOREST DESTRUCTION

Aogán Ó Rathaille's work (c1670 - c1729) provides a rare poetic insight to a changing – mainly Munster – landscape. Born in *Screathan an Mhil* (Scrahana-veele), in the *Sliabh Luachra* area of Co. Kerry, his most celebrated poem is "*Gile na Gile*" ("Brightness Most Bright"). In "*Créachta Críoch Fódla*" ("The Wounds of Ireland") and "*An Milleadh D'imigh ar Mhór-Shleachtaibh na hÉireann*" ("The Ruin that Befell the Great Families of Ireland"), he depicts a landscape and a forest eco-system without hope of re-covery. The versions used here are translations by Michael Hart-nett, who captures the theme of relentless despair.

Ó Rathaille's poetry juxtaposes his own dislocation and Mun-ster's political upheaval that began with the Flight of the Earls.

He eulogises and later criticises the Jacobite Browne landlords, but his real allegiance is to the defeated and dispossessed McCarthys, a Gaelic Munster landholding family. Ó Rathaille tries in vain to keep the Bardic tradition alive but he is cut adrift without land, patronage or privilege. He not only weaves the woodland into this narrative, but also apportions blame for its destruction.

While the "English bleaters" top the list, some of his most vitriolic poetry is reserved for two local exploiters – Tim Cronin and Murtagh Griffin – who "got what they could by the destruction of the woods, or by the extor-tion of hearth-money [fireplace tax]", according to Fr. Patrick Dineen in *The Poems of Egan Ó'Rahilly* (1900).

approximately 1550 to 1850 – was one degree centigrade less than it is today in contrast to the four centuries 950 to 1250 or Medieval Warm Period. The Gaelic poetry during this warmer era contrasts icily with Ó Rathaille's cold, infernal landscape:

> *Her branches rotten, her forests leafless,*
> *the frosts of Heaven have killed her streams now.*

Jonathan Swift supports Ó Rathaille's view on forest de-cline. When visiting Tipperary in 1732, he describes the county as "like the rest of the whole kingdom, a bare face of nature without houses or plantations".

Ireland wasn't the only country in Europe that experienced forest decline but despite large-scale over-cutting, most Euro-pean countries preserved their forest continuum and wood cul-ture. Like Ireland, Europe's forests were in decline from the sixteenth century, but issues such as sustained forest yield were being addressed as far back as the seventeenth century in France, Germany and to a lesser degree, England. Publica-tions such as Jean-Baptiste Colbert's *French Forest Ordinance of 1669* and John Evelyn's *Sylva: Or a Discourse on Forest Trees and the Propagation of Timber* (1664) addressed the exploitation of woodlands while sustainable forest management or *Nachal-tigkeitsprinzip* was being practised in Germany by the late eight-eenth century. There are no similar initiatives in Ireland.

FOREST DECLINE PARTIALLY ARRESTED

By the end of the nineteenth century the restoration of Euro-pean forests was underway. By the middle of the twentieth cen-tury, forests covered 24% of the land area of Europe, compared with 2% in Ireland and 5% in Britain. This decline was evident even in Co. Wicklow which had above average forest cover, as outlined by Michael Carey. One initiative that partially stemmed the flow of forest decline in Ireland was a planting scheme promoted by the Dublin Society – now the Royal Dublin Society (RDS) – in the 1740s. Part-funded by the Irish Parlia-ment in Dublin from 1761 and administered by the Society, it aimed to establish plantations and arrest the dramatic decline of Irish forests.

The scheme was the first private-State afforestation part-nership in the world. Despite the withdrawal of funding after the Act of Union came into effect in 1801, planting continued but was almost negligible by 1881 when the Land Law (Ireland) Act was introduced. The scheme supported a wide range of species, both native and introduced. While native oak, ash, wych elm and reintroduced Scots pine were planted, there was a heavy reliance on mainly European broadleaves including sweet chestnut, beech, English elm and walnut. In 1760 the list was extended to include European conifers including Norway spruce, silver fir and European larch, while Weymouth pine, a native of eastern North America was also favoured. In the 1780s the Society introduced unconventional species such as Athenian poplar, swamp pine and two-thorned acacia, although it appears that growers wisely avoided this particular species experiment. The scheme was later extended during its life to include planting by tenant farmers. The Society also published Samuel Hayes's *A Practical Treatise on Planting and the Management of Woods and Coppices* in 1794

The scheme was well underway when Arthur Young first visited Ireland in 1776. Young discussed the decline of Irish forests with landlords, some of whom "laid the destruction of timber to the common people". Young was not impressed with this view and is scathing when apportioning blame for the demise of Irish forests as he outlined in *A Tour in Ireland, 1776-1779*:

> The profligate, prodigal, worthless landowner cuts down his acres and leaves them unfenced against cattle, and then has the impudence to charge the scarcity of trees to the walking-sticks of the poor, goes into the House of Commons and votes for an Act, which lays a penalty of forty shillings on any poor man having a twig in his possession which he cannot account for. This Act, and twenty more in the same spirit, stands at present as a monument of their self condemnation and oppression. They have made wood so scarce, that the wretched cottars cannot procure enough for their necessary consumption; and then they pass penal laws on their stealing, or even possessing, what it is impossible for them to buy.

However, Young acknowledged good forestry practice during his tour when he visited Adare Manor, Co. Limerick, Monivea, Co. Galway and especially Castle Caldwell, Co. Fermanagh. The RDS scheme did succeed in arresting forest decline if only temporarily. William J. Smyth points to the 1841 census, which:

> … records the great surge of tree-planting between 1791 and 1841. The acreage under woodlands and trees is judged to have increased from 143,000 acres in 1791 to close on 500,000 acres by 1841. There was therefore at least three and a half times the number of planted trees in Ireland by 1841 than existed a half-century earlier.

While the Society deserves praise for the scheme, its legacy is downplayed in a report delivered to the Departmental Committee on Irish Forestry in 1908 by Richard Moss:

> The [premium] system seems to have done a great deal indirectly and by force of example. It did not, however, lead to the creation of a single plantation on a really large scale, and it can scarcely be claimed that it promoted forestry in the proper sense of the term.

By the mid nineteenth century, circumstances well outside the control of the Dublin Society planting scheme sounded a death knell for forestry that would last for 70 years. Sustainable forest development requires social, political and economic stability, all of which were lacking from the 1840s. The population

Arthur Young was extremely critical of landlords in Ireland who mismanaged their woodlands. In A Tour in Ireland, 1776-1779, *he identified a few exceptions where there was evidence of good forestry practice including Adare Manor, Co. Limerick (opposite), Monivea, Co. Galway (above) and Castle Caldwell, Co. Fermanagh (above right).*

had more than doubled to 8.2 million between 1785 and 1841 and within a few years the country was devastated by the Great Famine (1845-51). Afforestation had now virtually ceased as a climate of upheaval and major unrest followed, including increased demands for land reform and self determination.

Beginning in 1870, a series of Land Acts allowed the transfer of land from landlords to tenant farmers. Departing landlords cashed in their timber crops while tenant farmers, who now began purchasing their holdings concentrated on food rather than wood production. In 1902, Dr. Robert T. Cooper founded the Irish Forestry Society to impress "upon the British government the urgent need that exists for the rectification of the disastrous neglected condition" of forestry in Ireland. The Society played a part in the setting up of a state forestry department as outlined by Hugh Crawford but forest decline continued up until the Society's disbandment in 1923.

STATE FORESTRY

State intervention in forestry began the year after the devastating storm of 1903, described by James Joyce in *Ulysses* as both a "cyclone" and "the big wind of last February, a year that did havoc the land so pitifully". W.B. Yeats was shocked by the damage to Coole Park woodland where the storm. "blew down so many trees and troubled the wild creatures and changed the look of things ...".

The devastation caused by the storm and the continuing reckless exploitation of woodlands reduced Ireland's forest cover to little more than 1% of the land area. This influenced the then Department of Agriculture and Technical Instruction (DATI) to commission Professor John Nisbet of the West of Scotland Agriculture College to report on "plantations and waste-lands in the southeast" which were damaged by the storm. Nisbet understood that a long-term investment such as forestry held few attractions for farmers who manage seasonal crops, with seasonal incomes. Even on marginal land, the emphasis was on food rather than wood. This was understandable in a country which had lost its forest culture and where memories of the Great Famine were deeply ingrained in the popular imagination. Nisbet called for State intervention in developing a long-term forestry strategy: "If this be a duty at all, it is the duty of the State and not of the private landowners."

The following year, the State acquired Avondale House and estate in Co. Wicklow, along with bare mountain land for planting, to begin a new chapter of State forestry in Ireland. Avondale, the ancestral home of Charles Stewart Parnell, was converted to a school for trainee foresters. In the following years, 100 different tree species were established on the wooded estate to assess their suitability as forest trees in Ireland.

The purchase of partially wooded estates and bare land throughout Ireland soon followed. The aim was to conserve and reforest the few remaining estate woodlands and begin an ambitious State afforestation programme.

Forestry was widely regarded as the land use of last resort so only the poorest of land – mainly bare mountain and bog land – was made available for new planting. Caught between a rock and a soft place, so to speak, foresters were faced with a serious silvicultural dilemma. They had seen the results of two previous disastrous planting experiments with native and European species in bogs and uplands in counties Wicklow and Galway (see Chapter Seven). These experiments demonstrated that native and European species would, at best, survive but productive forests were required in a country which relied almost exclusively on timber imports. Augustine Henry, the botanist and forester, believed that Ireland's limited indigenous species menu needed to be supplemented by species from countries with similar climatic conditions to Ireland. The performance of exotic tree species trialled in Avondale from 1908 indicated that a number of these would perform well not only in Ireland's climatic conditions but also on the poor land which was made available for forestry during most of the twentieth century. As a result, the shift to European conifers, which began in the eighteenth century Dublin Society forestry scheme, accelerated to include conifers from regions identified by Henry. Species selection now included Sitka spruce, lodgepole pine and Douglas fir from western North America and Japanese larch, a native of Honshu Island, Japan.

By 1950, when only 2.1% of the land was under forest a major afforestation programme was launched by the first Inter-Party Government, formed in 1948. Ironically it wasn't initiated by either minister of the two relevant government departments – Agriculture and Lands – but by Seán MacBride, Minister for External Affairs and later winner of the Nobel Peace Prize. As a result, the Government adopted a long-term economic plan which included an annual planting programme of 10,000ha, eventually achieved and surpassed from 1959 to 1964. Small wonder, the Society of Irish Foresters awarded the Nobel Laureate, honorary life membership in 1984. The Society, founded in 1942, did much to advance and spread the knowledge of forestry as well as representing professional foresters, a function it still carries out.

The State relinquished its operational hold on Irish forestry in the 1990s and handed the baton to private landowners, mainly farmers. This coincided with the establishment of Coillte as a State commercial forestry company, which began trading in 1989. Coillte received a commercial mandate to manage most of the existing State forests with a small percentage of semi-natural woodlands being transferred to the National Parks and Wildlife Service. Within three decades, private ownership of the national forest estate had increased from 20% to almost 50%. Based on afforestation forecasts Ireland will eventually achieve the current European ratio of 60% private and 40% public forest ownership.

Average annual planting programmes by the State increased from 700ha in the 1920s to 2,100ha in the 1930s. The programme was divided between afforestation – planting bare land – and reforestation – planting harvested forests. Major State afforestation programmes were carried out from the 1950s until the early 1980s. Planting blanket peats such as Castledaly Forest Co.Galway, 1960 (above), was discontinued in the early 1990s when better quality land was made available for afforestation. At this stage most private planting was carried out by private forest owners, mainly farmers.

Looking back over the millennia, it is far too easy to simplify the Irish forestry narrative. Colonists are always the first target because they have historic form when it comes to the destruction of natural resources and Ireland is no different in this regard. A century after independence, it is now time to look at more recent history; to take stock of what Ireland has achieved in restoring a lost resource and the actors involved in this restoration. Those who criticise Augustine Henry and Irish foresters conveniently ignore the perilous state of forestry for most of the twentieth century and the poor quality land that was made available to create a forest resource. Faced with the inhospitable environment of the early planting programmes, foresters and forest workers created a resource that plays a major role in rural development, climate change mitigation and recreation. The recreational value of this major resource is appreciated by the Irish public and tourists who make 38 million visits to Irish forests annually, comprising 29 million visits in the south – as estimated by Mary Ryan and fellow authors – and 8.97 million to Northern Ireland Forest Service forests, according to Cognisense. The public's appreciation of this amenity was demonstrated in 2012 when government plans to sell Coillte's harvesting rights emerged. Nationwide meetings were organised as protesters feared that public access would be denied to Coillte's 400,000-ha forest estate including its 269 recreational sites. An open forest policy which has been in place since the 1970s is continued by Coillte, as well as woodlands managed by the National Parks and Wildlife Service and a number of local authorities with parks and woodlands. The protests were successful and the plan to privatise the State-owned company was dropped.

The move to private afforestation has created a number of opportunities and challenges. The involvement of private landowners, mainly farmers, in afforestation has resulted in a more diverse range of species as better quality marginal land is now becoming available for planting. Broadleaf planting, which comprised 5% of total planting for most of the last century under State afforestation, increased to an average of 37% between 2008 and 2012 when ash dieback disease caused farmers to temporarily revise their species selection.

While there remains a reluctance by some farmers to plant even marginal land that is eminently suitable for forestry, attitudes are changing. There is a growing acceptance of trees and woodlands as a land use that complements rather than competes with agriculture. There is also an increasing realisation that forestry can play a major role in helping farmers achieve climate change mitigation targets in greenhouse gas reduction. While private investors have stepped in to replace the State's role in afforestation, most new planting is carried out by farmers. Regardless of the level of private investment, the achievement of the government's target of 17% forest cover depends on farmers' commitment and willingness to engage in planting and maintaining the forests of the future. This will

require a partnership approach between the State and farmers; whereby the State provides financial support to assist the farmers through the establishment and non-productive phase of the forest with the farmer committing to forest management for generations to come, long after the State has relinquished its role and responsibility.

How best to achieve a partnership approach to ensure a viable forest estate is explored in Chapter 9. Government afforestation support, administered by the Forest Service, favours farmer planting while initiatives including the Native Woodland and NeighbourWood schemes encourage both farmers and community groups to establish and manage farm and urban woodlands.

The purpose of forestry has undergone a dramatic re-evaluation in recent years. While productive forestry is likely to continue to be the cornerstone of forestry policy, there is little doubt that multipurpose – including non-wood – forestry will

The first group of forestry students began training in Avondale, Co. Wicklow in 1904 after the former home and estate of Charles Stewart Parnell was purchased by the State. Over a century later, a series of nationwide protests was organised when the sale of Coillte's harvesting rights was mooted in 2012 as protesters feared public access to State forests would be denied. Shortly after a major protest in Avondale, in 2013 (opposite), the Government's privatisation plans were dropped.

receive far greater prominence in the future. However, multi-purpose forestry is not easy: it has to satisfy the needs of society while not losing sight of the commercial objectives of owners and the people who work the forests.

Regardless of objectives and silvicultural systems, sustainable forest management will be central to the future forestry paradigm. Sustainable forestry maintains the forest continuum, with benefits for all our people: those who visit the forest; those who work the forest and those employed in downstream industries who depend on renewable wood for construction, furniture, energy and countless other uses, just as they did since the arrival of first people to the island 12,500 years ago. The major difference this time is that the forest is allowed to renew itself and play a positive role in climate change mitigation and sustainable living.

Across Earth's ecosystems, wildfires are growing in intensity and spreading in range. From Australia to Canada, the United States to China, across Europe and the Amazon, wildfires are wreaking havoc on the environment, wildlife, human health, and infrastructure.
– Spreading like Wildfire – The Rising Threat of Extraordinary Landscape Fires. 2022. *UN Environment Programme.*

CHAPTER THREE

Forest Protection: Local and Global

No bit of the natural world is more valuable or more vulnerable than the tree bit. Nothing is more like ourselves, standing upright, caught between heaven and earth, frail at the extremities yet strong at the central trunk; and nothing is closer to us at the beginning and at the end, providing the timber boards that frame both the cradle and the coffin.
Seamus Heaney, *Sculpture in Woodland*, 2004

Forests are the planet's greatest renewable resource if sustainably managed. If mismanaged, they lose their capacity to renew. They also lose their multipurpose function which provides wood and non-wood benefits. Over the millennia, the planet has lost most of its natural forests due to a combination of factors including population pressure, expansion of agriculture and over-exploitation for fuel, construction and other uses.

THE GLOBAL FOREST

Forest cover still amounts to 31% of the earth's land surface, so all is not lost. Global deforestation has slowed down in recent years, but this is due to forest expansion and sustainable development, especially in temperate forest regions (Figures 3.1 and 3.2). The threat to tropical forests remains, so the global imbalance of forest destruction needs to be corrected.

Because forestry is a multipurpose land use, deforestation causes a myriad of catastrophic effects that go beyond the obvious loss of wood for construction, energy, furniture, paper and hundreds of uses we take for granted. Ultimately, loss of forests results in loss of livelihoods, biodiversity, medicine sources, shelter, clean water and soils due to erosion and flooding.

According to the UN Food and Agriculture Organisation (FAO), "the world has lost a net area of 178 million ha of forest" from 1990 to 2020, which is more than the area of Ireland, France, Spain, Germany and the UK combined. On a more positive note the FAO claims the annual rate of net forest loss has declined in recent decades: 7.8 million ha (1990-2000); 5.2 million ha (2000-10); and 4.7 million ha (2010-20). These statistics

Forest fire in spruce and pine forest, Germany. Forest and wildfires are now a global problem, which is being exacerbated by global warming.

Figure 3.1: *Annual change in forest area 2010-2015 (1,000ha/year).*

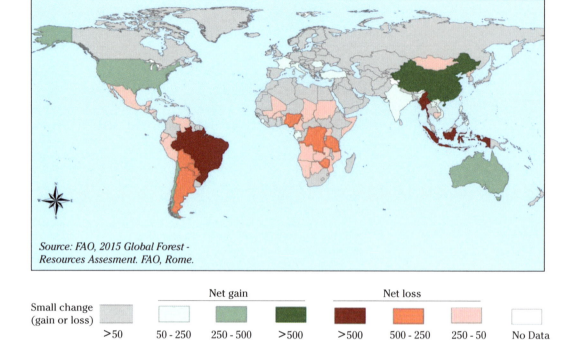

Source: FAO, 2015 Global Forest - Resources Assesment. FAO, Rome.

Small change (gain or loss)		Net gain			Net loss			No Data
>50	50 - 250	250 - 500	>500	>500	500 - 250	250 - 50		

Figure 3.2: *Forest area (% of land area under forest).*

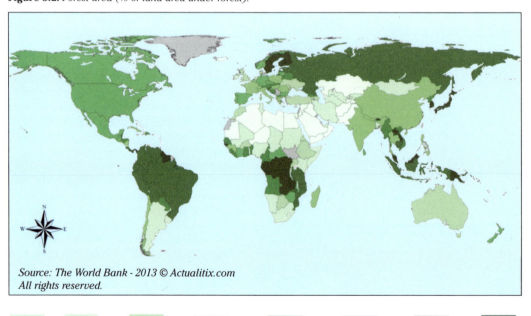

Source: The World Bank - 2013 © Actualitix.com All rights reserved.

>3.33	3.33 ; 11.13	11.13 ; 21.49	21.49 ; 32.71	32.71 ; 39.12	39.12 ; 48.77	48.77 ; 62.55	>62.55

THE GLOBAL FOREST

The following are the three major global forest biomes:

Tropical forests: Located in tropical and subtropical regions or within 23.5°north and south of the equator, they are also referred to as Rainforests because annual rainfall is more than 200 cm, while temperatures are generally between 20°C and 30°C. They have the highest species diversity in the forest biome. They have two seasons – rainy and dry – while the length of daylight is 12 hours with little variation.

Temperate forests: Located between 25° and 50° latitude in the northern and southern hemispheres, they are the predominant forests in western and central Europe, northeastern Asia and North America. Very few of the primeval temperate forests remain intact. They have four distinct seasons and temperature varies from -30°C to 30°C, although maximum and minimum temperatures are increasing due to global warming. They change from predominantly broadleaf to coniferous as they approach the boreal forests.

Boreal forests: The largest terrestrial biome, the boreal or taiga occupies territory between 50° and 60° north latitudes, comprising Eurasia and North America, most of Siberia, Scandinavia, Canada and Alaska. Seasons are divided into short, moist, and moderately warm summers and long, cold winters. Tree species are confined almost exclusively to conifers, which are threatened by climate change resulting in permafrost degradation according to a number of reports including a study by Katherine Dearborn and fellow researchers in 2020.

mask an underlying trend of deforestation especially in the tropics. The area of global forest decline would be much higher as countries in Europe, North America, parts of Asia, including China and India, have either increased or stabilised the level of forest cover in recent years (Figure 3.3). Large-scale deforestation still occurs in the tropics, especially Africa and South America, regions that also experience poverty, political uncertainty and the worst effects of global warming.

Agricultural expansion continues to be the main driver of deforestation, forest degradation, and the associated loss of forest biodiversity. Large-scale commercial agriculture (primarily cattle ranching and cultivation of soya bean and oil palm) accounted for 40% of tropical deforestation between 2000 and 2010, and local subsistence agriculture for another 33%, according to the FAO. While the global forest remains under serious threat, there are positive signs that deforestation is being addressed, albeit slowly as identified by the FAO:

> There is quantitative evidence to show that forests are being managed more sustainably and that forests and trees contribute to achieving sustainable development goals relating to livelihoods and food security for many rural poor [with] access to affordable energy, sustainable economic growth and employment (in the formal sector), sustainable consumption and production, and climate change mitigation, as well as sustainable forest management.

The report outlines the importance of forests and trees that "provide around 20% of income for rural households in developing countries, both through cash income and by meeting subsistence needs". It outlines the importance of maintaining the equilibrium of multipurpose forestry. While wood is an essential component of the forest, non-wood products and

Figure 3.3: *Net forest change by region, 1990-2020 (million ha, annually)*

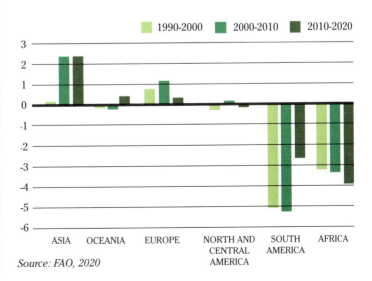

Source: FAO, 2020

services can be just as important in providing "food, income, and nutritional diversity for an estimated one in five people around the world, notably women, children, landless farmers and others in vulnerable situations".

Water is an essential part of the non-wood mix in sustainable forest management. Water retention by forests plays a major role in flood prevention, but other factors have changed the dynamic between forests and flooding including flood plain management as outlined by the FAO:

> Water quality, essential to the health and life of both rural and urban populations, is directly related to forest management. Changes in land cover, use and management have grave implications on a nation's water supply. While three-quarters of the globe's accessible freshwater comes from forested watersheds, research shows that 40 percent of the world's 230 major watersheds have lost more than half of their original tree cover. Despite this, the area of forests managed for soil and water conservation has increased globally over the past 25 years, and in 2015 a quarter of forests were managed with soil and/or water conservation as an objective.

Achieving a balance between food and wood is a major challenge. The FAO reports "Trees have been cut down in the forest north of Chiang Mai, Thailand in order to make new fields for agriculture".

Loss of forests through overcutting and burning, results not only in loss of carbon sinks but also increases in greenhouse gas as the carbon stored in the destroyed trees is emitted back to the atmosphere. Climate change has generated, disease and windthrow as well as droughts that allows wildfires and forest fires to burn longer as evidenced in Australia and Brazil in recent years. So, even at a time when forest clearances are falling, green house gas reduction emission gains are wiped out by increases in wildfires.

This is particularly true of the Amazon Forest, regarded as "the lungs of the planet". In 2019, 87,000 forest fires were recorded in the Amazon Forest in Brazil. While this destruction was much less than the average number of fires recorded during the period 2002-2007 (Figure 4), reports of the

Figure 3.4: *Total number of fires in Brazil (1 January-29 August) from 1998 to 2019.*

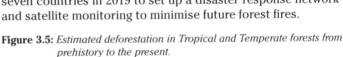

Source: FAO. The State of the World's Forests. Food and Agriculture Organization of the United Nations. Rome, 2012. *Based on estimates published by the FAO and Williams, M. (2002)* – Deforesting the earth: from prehistory to global crisis.
www.ncbi.nlm.nih.gov/pmc/articles/PMC5852684/

deliberate burning of large tracts of Amazonia in 2019 as part of government policy to make way for grazing and other land uses is a particularly disturbing development.

Brazil has a central role to play in protecting the Amazon forest. While 70% of Amazonia lies within Brazil's borders the Rainforest stretches into Bolivia, Colombia, Ecuador, Guyana, Peru and Suriname, where fires have also destroyed natural forests. A positive development was the signing of a pact by all seven countries in 2019 to set up a disaster response network and satellite monitoring to minimise future forest fires.

Figure 3.5: *Estimated deforestation in Tropical and Temperate forests from prehistory to the present.*

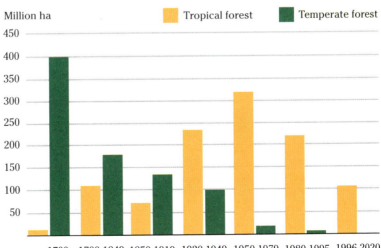

But it's not just the Amazon that battles against deforestation. In 2019, some 16 million ha of forest were burned in Russia according to the country's Federal Forestry Agency, which Greenpeace estimates at 500 million tonnes of CO_2, as much as Saudi Arabia's annual greenhouse gas emissions. Greenpeace maintains that Indonesia is the world's fourth-largest emitter of greenhouse gases, mainly due to "peatland fires and ongoing deforestation mostly to produce cheap commodities such as palm oil".

Large-scale deforestation still occurs in the tropics especially in regions that also experience poverty, political uncertainty and global warming.

While we tend to focus on countries which experience intense dry periods, increased forest and wildfires are now a global problem, exacerbated by global warming. The UN Environment Programme (2022) outlined the geographical threat of fire:

> Across Earth's ecosystems, wildfires are growing in intensity and spreading in range. From Australia to Canada, the United States to China, across Europe and the Amazon, wildfires are wreaking havoc on the environment, wildlife, human health, and infrastructure.

The report says we must learn to live with fire but we must learn to better manage and mitigate the risk of wildfires to human health and livelihoods, biodiversity, and the global climate. This includes countries like Ireland. While we don't experience catastrophic fires, we cannot take this for granted in future. For example Arctic areas, which never experienced major fires, are now at risk. Regions such as northern Siberia, will experience drying conditions that are "likely to increase burning by the end of the century" according to the UN report.

PROTECTING EUROPE'S FORESTS

Europe's past experience in neglecting and overexploiting its forests leaves us in no position to lecture developing countries about deforestation. Neither, does it allow us the luxury of looking the other way, as deforestation is a global issue with global consequences. Historically, Europe and other developed countries with temperate forests destroyed their forests at a rate that matched the destruction of tropical forests today (Figure 3.5). And worse, we continued up until recent times to facilitate illegal logging as we availed of cheap wood and wood products from developing countries. The EU faced up to its responsibilities in addressing global deforestation, especially illegal logging, when Member States adopted the Forest Law Enforcement, Governance and Trade (FLEGT) Action Plan in 2003 which is being updated (after this book goes to print) by a new European Deforestation Regulation (EUDR). Europe, along with developed regions, had earlier introduced forest certification to encourage countries around the world – especially in the tropics – to practise sustainable forest management (SFM). FLEGT's aim is "to reduce illegal logging by strengthening sustainable and legal forest management, improving governance and promoting trade in legally produced timber". A key aspect of FLEGT is the creation of Voluntary Partnership Agreements (VPAs), legally binding trade agreements between the EU and timber-producing countries whose forests are threatened due to illegal logging. The purpose of VPAs is to ensure that timber and timber products exported to the EU are sourced legally and from sustainably managed forests. The agreements also help timber-exporting countries stop illegal logging by improving their forestry regulation and governance.

Forest certification has been an important development in tackling illegal logging. Certification is independently verified proof that wood and wood products originated from sustainably managed forests (see Chapter 4). Certification originally focused on tropical forests, but countries in the developed world, including Europe, decided to lead the way.

The involvement of Europe in addressing the threat to tropical forests is also an acknowledgement of the threat that global warming brings to temperate and other forests. While human exploitation is the main cause of global forest destruction, the threat to the forest resource is no longer confined to developing countries in the tropics. Forest destruction now exists even in countries that practise sustainable forest management.

Threats to European forests
Reports from organisations such as the European Forest Institute (EFI) illustrate the damage caused to European forests in recent years, which "has increased markedly". Disturbance can be either abiotic, biotic or a combination of both. Abiotic refers to storm, fires and drought damage while biotic includes fungal disease, insect infestation and animal damage. Abiotic and biotic can combine as evidenced during Cyclone Friederike and its aftermath in Central Europe in early 2018. While the cyclone was a central factor in forest destruction, drought exacerbated the damage as it was followed by insect damage and in some instances, fire.

Most of the disturbance was caused in Germany where some 17 million m³ of timber was blown while damage in Austria, France, Benelux and northern Italy amounted to an

estimated 10 million m³. This timber needed to be harvested quickly to minimise the ensuing damage caused by the eight-toothed spruce bark beetle or *Ips typographus*. For example, 18 million m³ of timber in Czech Republic forests was damaged by bark beetle infestation – 10 times the normal annual damage. EFI acknowledges that it is difficult to statistically prove trends in frequency and severity of this damage, because data series are limited to a few hundred years but in the case of forest windthrow, the authors claim "there is evidence that the actual severity of storms in the wake of climatic changes may increase during the next decades".

Threats to Irish forests
Biotic and abiotic threats are lower in Ireland than other European countries but there have been worrying trends of increased activity. On a positive note, while fires can destroy over 1,500ha during a season, the annual average area of forests damaged by fire in Ireland during the 20-year period, 1999-2018, was 540ha. This is less than 0.1% of the forest estate every year. Wind is the most significant abiotic factor causing damage to forests in Ireland and Europe. In Ireland, average annual storms damage 15% of the annual harvest. Even in exceptional years such as 2014, when Storm Darwin flattened over 2 million m³ of timber, this was approximately half the annual harvest or a quarter of the annual gross volume increment of Irish forests for that year.

Irish forests have escaped catastrophic fire and wind damage, as well as serious attacks of insects such as the eight-toothed spruce bark beetle and the great spruce bark beetle (*Dendroctonus micans*). However, there are worrying biotic trends that may be irreversible. Our island status has been one of the reasons why Irish forests have remained largely free of insect and fungal disease damage but globalisation and climate change have increased the risk, especially of both bark beetles, which are present in Britain. Our open economy allows for potential imports of these and harmful fungal diseases.

Apart from Dutch elm disease which infected native wych elm and naturalised English elm in the late 1960s, other tree species are now at risk. Since the introduction of ash dieback in 2012 – caused by the fungus *Hymenoscyphus fraxineus* – most ash trees in woodlands and hedgerows have been wiped out throughout Ireland. Two years earlier, *Phytophthora ramorum* infected Japanese larch trees after it spread from imported ornamental rhododendron and viburnum to the wild. As a result, ash, wych elm and Japanese larch are no longer planted in Ireland. These important species may be lost to Irish forestry forever, unless tree breeding research programmes, to develop resistant species, are supported.

The threat to a number of tree species in Ireland and Europe is now real, although no longer due to overexploitation as natural forests are protected and new forests are being created. The main difference between past and present overexploitation

Norway spruce, Germany – forest infested by Ips typographus, *the eight-toothed spruce bark beetle, which has destroyed large tracts of forests in southern and central Europe. The beetle thrives in areas where trees are under stress due to drought which is now a feature of global warming. In addition, beetle damaged forests are prone to fire damage.*

is that we now understand the difference between sustainable and unsustainable development and the resultant local and global impact.

Balancing food and wood security

We also understand the challenges of achieving a balance between wood and food security; the need to protect and expand the depleted forest resource without damaging food production. When the restoration of Europe's forests began in the nineteenth century, the world's population was less than 1.5 billion so the increase in forest cover had little impact on land availability for food production, although famines occurred for much of the century. As the world's population is projected to increase from around 7.9 billion today to close to 9.5 billion by 2050, a major challenge of our time is to increase agricultural production and maintain food security while increasing forest cover and restoring our lost forest resource. In this regard, Ireland is the only country in Europe that can increase forest cover by at least 60% during the period 2021-2050 without negatively impacting on agricultural production. This is a key area where forestry can help agriculture to achieve demanding greenhouse gas reduction targets.

Against this background the answers to "Why forests? Why wood?" may seem self evident, especially for Ireland. Issues such as climate change, increases in greenhouse gas emissions, land use pressure and unsustainable development have highlighted the role of forests. The environmental, social and economic benefits in sustainably managing existing forests, creating new forests and increasing wood usage to replace fossil based materials are now central in EU climate change policy.

The arguments for new productive forests cannot be isolated from other land uses, especially agriculture. The balance between food and wood production is critical in a planet that

Logs salvaged after windthrow caused by Storm Darwin in Co. Kerry in 2014. On average, annual storms damage 15% of the yearly harvest in Ireland. Even in exceptional severe storm years, approximately half the annual harvest or quarter the annual gross volume increment of Irish forests are blown down which still amounts to only 0.1% of the total standing timber volume of 142 million m³.

will need to feed and house a population that will have quadrupled in a century. EU Horizon 2020 states: "70 % increase of the world food supply is estimated to be required to feed the 9 billion global population by 2050." The report outlines the challenges in balancing food production against climate change objectives as "agriculture accounts for about 10% [34% in Ireland] of the EU greenhouse gas emissions, and while declining in Europe, global emissions from agriculture alone are still projected to increase up to 20% by 2030".

The Earth's 4 billion ha of forests are the lungs of the planet but will not be enough on their own to sequester sufficient carbon or to provide wood for construction and other uses. The existing forests need to be nurtured and expanded where possible and natural forests conserved. But new forests need to be established because the planet needs more wood, which will require the planting of commercial forests.

Planted forests cover approximately 264 million ha or 7% of the global forest area. The area of plantation forests is increasing by 5 million ha annually. The FAO maintains that these forests "have the potential to provide two-thirds of the global industrial roundwood demand". In addition to their timber production benefits, well-managed planted forests provide social and environmental services "ranging from rehabilitation of degraded lands, combating desertification, soil and water protection, sequestering and storing carbon, recreation and landscape amenity".

Planted forest production capacity takes pressure off natural forests, allowing these the space and time to develop according to their own biodiversity requirements as is now happening in Ireland, where existing native woods are being conserved and new native woodland cover is being created with funding and support from the Native Woodland Scheme. This is compatible with the global approach as outlined by the FAO:

> Responsible management of planted forests can reduce pressure on indigenous forests for forest products and allow them to be designated for other protective and conservation purposes. They can also complement and supplement the REDD (reducing emissions from deforestation and forest degradation) and REDD-plus initiatives to reduce greenhouse gas emissions from deforestation and forest degradation in developing countries. As such, planted forests have multiple values, many of which cannot be provided by other types of land use.

David Attenborough's proposal is to conserve natural forests as well as farming trees to allow high yielding timber production on less land so that food and wood objectives are achieved and pressure taken off natural forests. Balancing the global need for a safe and secure food supply against climate change objectives is a major challenge of our time and David Attenborough's vision on the role trees and forests will play in our future planet has major local and global benefits.

SUSTAINABLE FOREST MANAGEMENT

SOCIAL ... wellbeing ... recreation ... viable employment ... renewable wood ... non-wood products ... carbon displacement ... energy ... flood control ... water quality ... soils ... biodiversity ... carbon sequestration ...

BIO ECONOMIC

ENVIRONMENTAL

CHAPTER FOUR

Sustainable Forest Management

Sustainable forest management … aims to maintain and enhance the economic, social and environmental value of all types of forests, for the benefit of present and future generations.
The Secretariat, United Nations Forum on Forests, 2009

Luckily for us, the solution to how we restore the planet's forests is simple. We just need to give trees the opportunity to draw on their natural resilience. We can start by protecting those last remaining ancient forests. These precious places, undamaged by humans, still host their entire natural mix of species, and trees young and old.
David Attenborough, *Our Planet: How to Save our Forests*, 2019.

Trees and forests are the greatest renewable resource on the planet, providing they are sustainably managed. Sustainable management ensures the forest is a renewable resource with the ability to sustain biodiversity, communities, economies and ultimately the planet. The importance of trees to the very survival of our planet is the most obvious answer to "Why forests?", but the question leads inevitably to how we protect and enhance a diminishing resource. Sustainability has to be at the heart of how we protect and nurture existing forests, and how we establish new forests. While forest ecosystems can be fragile, they can also be resilient if allowed the time and space "to draw on their natural resilience", as David Attenborough explains.

Sustainability is not a new concept in forest management. Sustained yield entered the European forester's lexicon in the late eighteenth century. It has been the guiding silvicultural principle in restoring Europe's forests from the nineteenth century after previous centuries of overcutting, not dissimilar to the illegal logging now taking place in many tropical countries. Sustained yield regulates the felling and regeneration of forests to ensure a continuous supply of timber to meet society's needs. The principle was introduced to Ireland in the early twentieth century, which was far too late to rescue the country's vastly depleted forests. Sustained yield management maintains the equilibrium between timber removed, retained

Figure 4.1: *Sustainable forest management is the wise stewardship of forests to meet social, bioeconomic and environmental objectives. These three pillars of sustainable management are subdivided into their constituent parts: from forest production to protection; and environment enhancement to employment and wellbeing.*

and reforested, so it ensures the forest continuum. In practice, forest volume increment exceeds the volume harvested, so in the event of future excessive timber demand or forest damage, there are sufficient wood reserves in place. For example, in Ireland net annual volume increment (the annual volume increase in forest growth after natural losses) in 2015 was 7.7 million m³ while only 5.0 million m³ of wood was harvested as estimated in *Ireland's National Forest Inventory* 2017. In Ireland, therefore 65% of the net annual increment is harvested compared with 73% in Europe according to *State of Europe's Forest 2020*.

Sustained yield is vital to ensure that the forest is a renewable timber resource. But more is expected from the forest as an ecosystem. As a result, sustained yield has been broadened to incorporate sustainable forest management or multipurpose forestry. This is a more expansive form of silviculture; more in tune with forest biodiversity.

Sustainable forest management was introduced in the late twentieth century to cater not only for production forestry but also for a range of other ecological and societal needs. Essentially, it is the wise stewardship of forests to meet social, economic and environmental objectives, often referred to as the three pillars of sustainable forest management. These are subdivided into their constituent parts: from forest production to protection; and environment enhancement to employment and wellbeing (Figure 4.1). The achievement of sustainable forest management in practice requires adherence to strict forestry environment guidelines (opposite panel).

In aiming to achieve this relatively new forestry paradigm, sustainable forest management also acknowledges the role – and voice – of stakeholders who value multipurpose forestry. In Ireland, it not only requires consultation with the 12,000 people who work in the forests and downstream industries, but also the increasing number of people who make 38 million visits to Irish forests – nine million in the north and 29 million in the Republic – every year, according to DAFM and DAERA data.

Globally, it includes 1.6 billion people whose livelihoods depend on forests and the 300 million who live in forests. They are the people who will ensure the future of their forests while their forests, in turn, will ensure their right to a healthy ecosystem as well as renewable products and sustainable livelihoods.

Multipurpose forestry endeavours to achieve a balance between wood and non-wood management objectives. This balancing act contains inherent pressures like any land use that ultimately looks to the marketplace to trade its produce. Timber trade has been the real test of sustainable management as deforestation remains a threat to the global forest resource, especially in tropical regions. Developed countries – mindful of their own historic overcutting – have provided the lead by adopting sustainable forest management. While challenging, sustainable forestry is well within the remit of foresters, forest

ENVIRONMENTAL GUIDELINES

In Ireland, forest operations throughout the growing cycle – from planting to final harvest – comply with procedures outlined in the Department of Agriculture, Food and the Marine Forestry Standards Manual. The manual sets out the procedures and operational standards that conform with sustainable forest management. In addition, forest owners are legally required to carry out operations in accordance with the conditions in the following forest environmental guidelines on:

- Water quality
- Landscape
- Archaeology
- Forest biodiversity
- Harvesting and environment
- Forest protection

These guidelines ensure the environmental aspects of sustainable forest management are implemented. Adherence to the guidelines is a condition of grant aid and the issuing of forestry licences. Forest operations such as planting, road building and harvesting are also screened on their potential impact on Natura 2000 sites within a 15-km radius or 700-km² area. Natura 2000 sites comprise Special Areas of Conservation (SACs) and Special Protection Areas (SPAs), designated to afford protection to the most vulnerable habitats and species in Ireland.

To ensure greater species diversity, all forests established under the Forestry Programme (2023-2027) must include at least 35% native species and open biodiverse areas. Forest establishment also is prohibited in threatened bird habitats in particular waders. For example afforestation cannot take place within 1.5km of curlew nesting sites (see also Chapter 5).

owners and other stakeholders. It requires an understanding and an acceptance by all stakeholders that in the ongoing forestry debate no single forest type can fulfil all expectations. As stakeholders have different expectations from forestry, sustainability has to be in some way measurable in order to achieve a balance between the social, environmental and economic goals of the forest.

FOREST CERTIFICATION

The accepted way of measuring the implementation of sustainable forest management to the satisfaction of stakeholders is

Deadwood is an important resource for biodiversity and soil amelioration in forests. Comprising dead trees, stumps and branches, the total deadwood volume in Irish forests is estimated at over 10.5 million m³.

independently verified forest certification. This is a voluntary process involving consultation between members of the public and forest owners. Timber processors and manufacturers who utilise wood and wood products are also part of the process which ensures chain of custody certification. Certification creates a better understanding of multipurpose forestry where non-wood aspects of the forest enjoyed by the public are

FOREST CERTIFICATION

The main reason for the introduction of forest certification is to encourage countries to practise sustainable forest management – especially in the tropics, where deforestation continues. It was introduced by non-governmental bodies in the 1990s as a means of tackling illegal logging, deliberate burning and forest degradation. However, developed countries led by example and have adopted certification in forest management practice. Certification is now increasing throughout Europe and developed countries.

Approximately 440 million ha of forest land around the world have been certified by the Programme for the Endorsement of Forest Certification (PEFC) and the Forest Stewardship Council (FSC) according to *The Global Forest Atlas*. While this represents only 10.7% of the total global forest area of 4.03 billion ha, it amounts to 29.6% of annual global industrial roundwood production, estimated at 523 million m³. In Ireland and most of Europe, certification is endorsed by PEFC and FSC.

Certification doesn't stop at the forest gate but applies throughout the forest supply chain, including companies involved in timber processing, manufacturing, printing and other end uses. These receive chain of custody certification.

Clonad Wood, Co. Offaly: Consultation with a wide range of stakeholders is part of the process necessary to achieve forest certification. Clonad Wood is managed by the Irish Forest Unit Trust.

counterbalanced by a greater appreciation of the rights of those who work the forest to earn a livelihood from its produce. The criteria for certification include:

- Maintenance of the forest ecosystem, its health and vitality.
- Optimisation of the productive capacity of forest ecosystems.
- Conservation and maintenance of soil and water resources.
- Maintenance of the forest's contribution to global carbon cycles.
- Enhancement of long-term socio-economic benefits to meet the needs of societies.

Certification is now increasing globally (opposite panel) while in Ireland all State forests – Coillte and Northern Ireland Forest Service – and an increasing number of private forests are certified. Achieving certification is too costly for individual small woodland owners, who are mainly farmers. Some of these are now joining group schemes or cooperatives to provide economies of scale in achieving certification.

Sustainable forest management has changed the perception and actuality of forestry among those who own and manage public and private forests, as well as the public who use the forest. Regardless of how urbanised the global population has become, people still rely on the forest as the provider of a renewable resource, clean air, biodiversity and recreation. The global forest has shrunk, while the global population has increased to almost 8 billion. As a result, there is growing pressure to restore and increase the forest resource without negatively affecting food security. Achieving both food and wood security doesn't arise in Ireland where there is room to increase our low forest cover, without negatively impacting agricultural production.

In Ireland, sustainable forest management is sufficiently versatile to allow a range of forest types, both native and introduced, to evolve with space to allow commercial and conservation management to coexist. This mirrors Sir David Attenborough's approach to global forestry, which is to protect and enhance "those last remaining ancient forests" with the need "to farm trees" to build houses and displace fossil-based materials. David Attenborough's call to conserve our ancient forests is tempered with his realisation that "The natural forests can't provide all the wood that we need". This forest duality – conservation and production – is also central to forest policy in Ireland, where the challenge lies in expanding the productive forest resource in tandem with the protection and enhancement of our native woodland resource.

CHAPTER FIVE

Why Forests?

A mountain is valuable, not because it is high, but because it is wooded.
Japanese proverb

A forest is the finest thing in the world: it is the expression of nature in the highest form: it is so full of beauty and of variety.
Augustine Henry, letter to Evelyn Gleeson, 1899

… the woods and groves are cut down, for there is need of an endless amount of wood for timbers, machines, and the smelting of metals. And when the woods and groves are felled, then are exterminated the beasts and birds …
Georgius Agricola from *De Re Metallica,* 1556

The forest's recreational importance creates goodwill towards the forest as a public amenity, but its benefits are also rooted in the community, where 33,000 people profit from the forest, comprising 23,000 forest owners who have established forests since 1980 and 10,000 people who work in forests or downstream industries. By 2035, Forest Industries Ireland (FII) predicts the creation of an additional 6,000 rural jobs while an extra 15,000 landowners – mainly farmers – may have established forests.

W hy forests? The question should be redundant given the wide range of wood and non-wood benefits provided by forests and woodlands. The world needs forests now more than ever; for the role they play in climate change mitigation, environment enhancement, in the renewable goods they produce and for their untapped potential in the bioeconomy. Sustainability is at the core of the many reasons why we should protect the vastly depleted forest ecosystem and create new forests and woodlands.

There is no single answer to why we need forests, just as there never was a single reason for their decline since man first cleared the primeval forest for fuelwood, for agriculture and for shelter. Later, when they were threatened by over-exploitation for industry, plausible reasons were always advanced for mass tree clearances just as they are today. Georgius Agricola dismissed arguments that deforestation would accompany "the smelting of metals" in Europe in the sixteenth century. He also dismissed the aftermath of mining where "the inhabitants of these regions, on account of the devastation of their fields, woods, groves, brooks and rivers, [would] find great difficulty in procuring the necessaries of life …". His dismissal has echoes in today's burning of the Amazon rainforests as Brazils former President Jair Bolsonaro rejected arguments that the destruction of the rainforest has dire consequences for this unique ecosystem as well as for the future of the planet.

The importance of forests in sequestering CO_2 and cooling the planet is accepted as pivotal to climate change mitigation. The forest as a habitat to protect and nurture biodiversity is now acknowledged in countries that have learned the lessons of past over-exploitation, but are still being ignored where burning and illegal logging persist.

Yet, in Ireland, despite centuries of the greatest forest decline in Europe, a resistance to forestry persists. Forestry – especially afforestation – has to prove itself socially, economically and environmentally. This is partially understandable in a country where historically the emphasis was on food production for its own people, particularly after the Great Famine. Farmers will argue, with justification, that a modern and efficient agricultural sector in Ireland puts food on the table of 30 million people in Ireland and worldwide, mainly in the UK. This has led to agriculture being the largest emitter of greenhouse gas (GHG) in Ireland. Forestry is now identified by government and the EU as a major land use to offset and reduce GHG emissions in agriculture

But attitudes to forestry are changing; among farmers who have planted most of the forests in Ireland since the 1990s and among the public where the forest's role as a place of solace and recreation is increasingly recognised in a more urbanised society. But it's not just farmers who need convincing. Other stakeholders also need answers to "Why forests?" in a country that plans to increase forest cover to 18% of the land within three decades. If forestry is to be accepted as a multipurpose resource the arguments need to be restated, including its impact on:

- climate change;
- people and wellbeing;
- agriculture;
- landscape;
- biodiversity;
- water quality;
- public goods; and
- the rural economy

FORESTS AND CLIMATE CHANGE

Afforestation is the single largest land-based climate change mitigation measure available to Ireland.
– Climate Action Plan: Securing our Future, 2021

Forestry is the only land use that can make a major contribution in decarbonising the economy. The contribution of forestry to climate change mitigation occurs in the forest where trees act as carbon sinks and after harvest when carbon is stored in wood and wood products. CO_2 is removed – or

THE GLOBAL CARBON IMBALANCE

Reducing greenhouse gas (GHG) emissions to limit global warming to 1.5°C by 2050 is humankind's greatest environmental challenge. While other GHGs contribute to climate change, CO_2 is the most abundant caused by human activity. Reducing the carbon source is vital but will not be sufficient on its own, as the net annual increase of 17.5GtCO_2 (Table 5.1) is too large.

Carbon sinks will also need to be increased. The most natural and sustainable way to increase carbon sinks is to eliminate deforestation, increase forest cover and use more wood to displace fossil-based material in construction and renewable energy.

Table 5.1: *Annual global carbon balance in gigatonnes of CO_2.*

Emmissions	CO_2
Fossil fuel combustion	34.8
Deforestation	4.1
Total emissions	38.9
Absorption	
Seas and lakes	10.2
Forests, vegetation, soils	11.2
Total absorption	21.4
Excess in atmosphere*	**17.5**

*Source: Adapted by D. Magner from Global Carbon Project 2021. Emissions averaged 2011 - 2021. *Imbalance of 3% (-1.0 Gt CO_2/yr) recorded (Friedlingstein et al) between global sources and sinks. Note: Data provided for CO_2 is measured in its weight (gigatonnes (Gt)). A Gt equals one billion tonnes.*

Figure 5.1: *Global greenhouse gas emissions, 2013.*

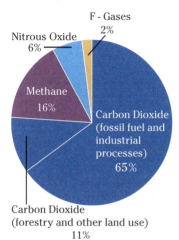

Carbon Dioxide
(forestry and other land use)
11%

Source: Intergovernmental Panel on Climate Change IPCC (2014).

Figure 5.2: *Ireland's greenhouse gas emissions, 2020*

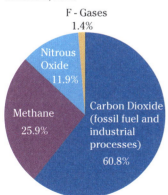

Source: Adapted by D. Magner from Environmental Protection Agency (EPA), Ireland (2021).

sequestered – from the atmosphere when it is absorbed by trees and other plants as part of the biological carbon cycle. While there are two other main greenhouse gasses (GHG) – methane (CH_4) and nitrous oxide (N_2O) – CO_2 is the greatest contributor to global warming. It is responsible for 76% of the total GHG emissions compared with CH_4 at 16% and N_2O at 6% (Figure 5.1). In Ireland, CO_2 levels are lower but because of a strong agricultural sector, CH_4 emissions are higher than the global average (Figure 5.2). CO_2 is released into the atmosphere through the burning of fossil fuels (coal, natural gas and oil) solid waste, biological materials including wood, and also as a result of certain chemical reactions such as the manufacture of cement and other fossil-based materials. David Attenborough sums up the vital role trees play in climate change and how we can use them to reverse the trend of global warming:

> Forests absorb carbon from our atmosphere and store it in their trunks, roots and the soil. They remove almost 15 [billion] tonnes of CO_2 each year. As we clear and burn forests, we release dangerous amounts of carbon back into the atmosphere, increasing the rate of climate change. We need to reverse this and create a world of expanding forests once more. Luckily for us, the solution to how we restore the planet's forests is simple. We just need to give trees the opportunity to draw on their natural resilience.

While climate change has taken millennia to evolve naturally, it has been accelerated by human intervention in a short period of time. The first and second industrial revolutions in Europe and the US, from around the mid-eighteenth to the early twentieth century triggered global warming as heavy industrialisation made demands on forestry and contributed to large-scale deforestation. In more recent times, the build-up of greenhouse gas has been increasing cumulatively, especially since the 1950s, from man-made sources including industry, energy, agriculture and – paradoxically – forestry. As a result of deforestation, the forest is losing its sequestration benefits and in some countries is a net contributor of greenhouse gas. One-sixth of global carbon emissions occur as a result of forest burning, overcutting and ecosystem degradation. But this situation can be reversed.

Sustainable forest management is key to the positive role trees play in climate change mitigation, as forests have the potential to absorb about 10% of the global carbon emissions projected for the first half of this century into their biomass and soils and products where they store them "in principle, in perpetuity" according to the FAO. In addition, wood which is used in construction locks in carbon during the life of the building as well as displacing fossil-based materials, while wood fuels are a renewable alternative to fossil fuels (as explored in Chapter 4).

Trees capture water, sunlight and soil nutrients naturally to produce renewable biomass. They do it more efficiently than any other natural or artificial entity. Wood biomass is produced

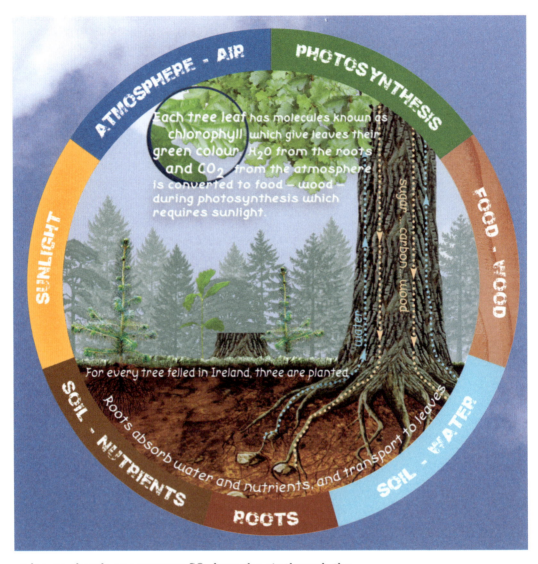

as leaves absorb or sequester CO_2 from the air through the complex chemical process known as photosynthesis. This process binds CO_2 with water and nutrients from the roots, and converts these into glucose, which is further transformed to wood. In this way, the carbon is stored inside the tree while oxygen is returned to the atmosphere (Figure 5.3). So, before the tree is required for timber, it is creating a cleaner environment by removing CO_2. In most of our forests, it performs this function without the aid of artificial fertilisation over the life of a crop rotation that may extend from 30 to over 100 years.

The process of carbon sequestration in trees ceases as soon as the tree is harvested or dies naturally. In managed forests, the trees are harvested but they will continue to store carbon in construction, furniture and other uses in the bioeconomy as well as displacing fossil-based material (as

Figure 5.3: *The forest carbon cycle*

Forests are the most efficient users of water, soil and sunlight to produce renewable biomass or wood. This is achieved only if harvested trees are replaced and habitats are protected. Forests play a key role in mitigating climate change by removing CO_2 from the atmosphere and converting it to carbon, which is then stored in the trees. How much CO_2 is stored in forests depends on yield, species and stocking but the following calculation provides an average answer based on wood volume and weight: Wood is composed of approximately 50% carbon, so on average $1m^3$ of dried wood weighing 500kg comprises 250kg carbon. When carbon is oxidised or transformed into CO_2, it creates 3.67 times its weight in CO_2 so 250kg of carbon creates 917kg – just under one tonne – of CO_2. This means that on average $1 m^3$ of wood stores 1 tonne of CO_2.

explored in Chapter 6). In protected forests where no harvesting is carried out, the forest contributes to greenhouse gas reduction. While the trees are left to die naturally and the stored carbon is returned to the atmosphere, natural regeneration of trees or replanting compensates for the loss of CO_2 while increased carbon will be locked in the soil.

Even though Irish forests cover only 11% of the land area, the total carbon reservoir or store currently exceeds 1 billion tonnes of CO_2, most of which is in the soil. Annual removal of CO_2 from the atmosphere exceeds 6.0 million tonnes or 3.6 million tonnes net of the CO_2 removed during timber harvesting. Maximising the long-term climate change benefits of Irish forests will require an annual national afforestation programme that includes short, to medium-term commercial rotations in conjunction with long-rotation mainly native broadleaves.

Ireland also acknowledged its international obligation in protecting and enhancing the global forests as one of the 145 signatories in the 2021 Glasgow leaders' declaration on forests and land use at COP26, which emphasised:

> the critical and interdependent roles of forests of all types, biodiversity and sustainable land use in enabling the world to meet its sustainable development goals; to help achieve a balance between anthropogenic greenhouse gas emissions and removal by sinks; to adapt to climate change; and to maintain other ecosystem services.

FORESTS AND PEOPLE

The interaction between the public and forests is regarded as one of the more important reasons why we need to protect our existing forests and where possible expand our forests and woodlands. These are proving to be vital places of recreation and wealth generation. Ireland's open state forest policy – north and south – is enjoyed by millions of Irish people and tourists while 33,000 people depend on the forest for its wealth and job creation benefits. These comprise 23,000 landowners and 10,000 workers in forests, in downstream industries and forestry-related services.

The importance of Irish forestry can be judged by the number of people who visit our forests as places of recreation, respite and renewal. Irish people and tourists make an estimated 29 million visits to recreation forests annually, while nine million visits are made to Northern Ireland's forests and forest parks. For visitors, the attraction of the forest is compelling, especially for an increasingly urbanised people who feel the need, more than ever, to communicate with nature.

This was demonstrated in 2012 and 2013 when there was large-scale public opposition to the proposed privatisation of State forestry. The proposal was eventually abandoned, but the experience provided government and public alike with an opportunity to reappraise the public good of the forest resource.

Hazel Wood, a popular recreation forest in Co. Sligo. It is one of almost 400 forests located around the island of Ireland, which are open to the public.

While forest cover in Ireland is low compared to other European countries, we are fortunate to have over 400 forests and forest parks which are open to the public throughout the island of Ireland, with additional woodland trails that people discover for themselves. The open forest policy adopted by Coillte and the Northern Ireland Forest Service, along with woodlands managed by the National Parks and Wildlife Service and other agencies demonstrates that social, economic and environmental objectives can be achieved by wise and visionary management. This approach points the way to the forests of the future; forests whose value is measured in their total contribution to society.

Forests are acknowledged not only for their physical health benefits but also as environments that enhance mental wellbeing. Research has shown that exposure to the natural environment reduces stress and anxiety, and among natural areas, forests are considered one of the more attractive as outlined by Yuki Iwata and fellow authors of the "Woodlands for Health" programme. They explore the health benefits that "woodland activities can provide to participants" and their "value to health professionals".

FORESTS AND AGRICULTURE

The global challenge of achieving a balance between food and wood security cannot be ignored in the climate change debate, especially in a planet where hunger is omnipresent. It is unlikely that the international community will ever agree on a global sustainable land-use policy, but there is room for sustainable forestry and agriculture to co-exist. As David Attenborough explains: "Around the world there are an estimated 2 billion ha of degraded land, where forests could be restored. That's twice the size of Europe."

In addition, forestry can be compatible with agriculture if developed on marginal or degraded land, with agriculture taking place on the better quality land. Both agriculture and forestry can be combined through the creation of agroforestry or silvopastoral systems. These allow grass cropping, livestock grazing and fruit production alongside conventional forestry.

The wood versus food debate doesn't apply to Ireland because of our low forest cover and sufficient non-productive agricultural land available for forestry. As a result, Ireland is the only country in Europe that can increase forest cover by approximately 65% without negatively affecting agriculture production. This would increase the area of forests to 1.2 million ha or 17% of the land area, still less than half the European average. This level of forest cover could be achieved if Ireland adopts annual afforestation programmes of 16,000ha as recommended by COFORD and 18,000ha as proposed by Dr David Styles and Prof. Cathal O'Donoghue in their University of Galway studies on achieving carbon neutrality by 2050.

RECREATION FOCUS

There are a number of State-supported publications and initiatives that encourage public participation in forest recreation. *Forest Recreation in Ireland – A Guide for Forest Owners and Managers* offers practical advice on developing woodlands and forests for recreation. It is aimed at forest owners and managers who are keen to develop recreational opportunities as well as community, sporting and environmental groups, local authorities and other statutory bodies.

The NeighbourWood Scheme is a forestry grant initiative aimed at promoting the recreational and wider social benefits of forests. The scheme is designed to create 'close-to-home' woodland amenities in partnership with communities for the benefit of local people.

These community woodlands vary in size and location, including forests in the countryside and urban outskirts such as Belleek Wood, Ballina, Co Mayo (above) and Newcastle-west, Co. Limerick (below).

Although agriculture is the highest emitter of greenhouse gas in Ireland, the sector's green international reputation is positive. It could be further enhanced by combining forestry with agriculture, especially on marginal farmland such as this mixed landscape in Co. Monaghan.

This would require transferring 2.5% of Ireland's land area to forest every decade. A land use change of this magnitude would need an even distribution of forests throughout the country, on suitable land, so that forestry could grow organically alongside agriculture.

Farmers who decide to shift some land use from unprofitable marginal agriculture to forestry can increase farming revenue as forestry provides a greater income (Figure 5.4) compared with most beef and all sheep systems, in addition to providing the added benefit of reducing greenhouse gas emissions in agriculture. In this scenario, forestry is compatible with green agriculture.

Figure 5.4: *Comparable average gross margin (€/ha) for six farm and two forestry systems (excluding Single Farm Payment, based on 2014 data).*

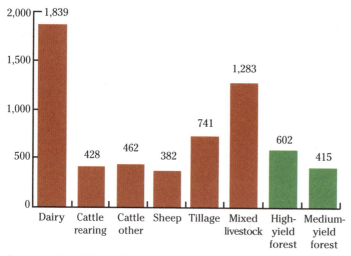

Source: *Adapted by D. Magner from Farm System Gross Margin Analysis, Teagasc National Farm Survey (A. Kinsella), and Teagasc FIVE Model (2015).*
Note: *"Cattle rearing" refers to predominantly suckler cow production while "cattle others" refers to mainly cattle fattening. "High-yield forest" is based on Sitka spruce yield class (YC) 24 (m³/ha/annum) while "medium-yield forest" is YC16. Income for forestry is based on annual equivalent value (AEV) or the future annualised value of the timber crop in today's money.*

Although, agriculture is the highest emitter of greenhouse gas in Ireland, the sector's green international reputation is positive. For example, the EPA Ireland Environment 2012 report refers to an EU study, which shows "that using a full life-cycle approach, Ireland's extensive grass-based systems produces the lowest greenhouse gas emissions in the EU for dairy animals and the fifth lowest for beef". The performance of the beef sector in emitting greenhouse gas varies considerably: unprofitable beef systems are the highest greenhouse gas emitters per kilogramme of live-weight of beef produced – up to 16kg of CO_2 equivalent per kilogramme of beef – compared with 8-11kg of CO_2 equivalent per kilogramme of beef produced in more efficient cattle farms. Professor Alan Matthews sums up the dilemma facing the Irish beef and dairy sectors:

> While the top third of beef farms by economic performance had a significantly positive net margin [in greenhouse gas emissions], the bottom performers have a significant negative net margin ...Those beef farms that are losing money are also the farms that contribute the highest emissions per unit of beef produced.

Reducing the number of suckler cows (cows reared for beef rather than dairy production) is a contentious issue. It presents political and social challenges, as forestry is the alternative land use on many of these farms. But reducing the number of suckler cows should not mean reducing the number of farmers if forestry is accepted as a land use that is compatible – rather than competing – with farming.

The economic argument alone will not convince farmers to transfer land from agriculture to forestry. Issues such as the long-term nature of forestry; the legal obligation to maintain land under forestry in perpetuity; landscape values; the compatibility of forestry and agriculture; and the geographic spread of forestry with its strong concentration on a few counties have been raised by farmers and their representative organisations such as the Irish Farmers Association (IFA).

In addition, social issues such as the stigma that planting – even marginal agricultural land – represents failure in farming terms is unique to Ireland. Large-scale afforestation programmes carried out in southern Sweden at the beginning of the last century have built the largest forest industry in the world. Closer to home, Scotland is creating a major forestry and forest products sector by adopting viable afforestation programmes. At worst, forestry in Ireland is regarded as the land use of last resort or at best an alternative land use. Martin Lowery, former CEO, Coillte, dismissed this systemic approach to forestry in Ireland:

> The term "alternative" enterprise, which is often used in relation to forestry, conveys an image of marginal activity which should be tolerated in the absence of anything better. This seriously underrates the value of forestry as the engine for development of many rural communities.

The large-scale State planting programmes carried out for most of the twentieth century continue to serve their purpose in providing timber for a viable timber-processing sector but a different template is required in farm forestry.

These programmes provided continuity in timber supply as continuity of commitment by successive governments achieved viable afforestation programmes. This has been lacking in the private sector. Continuity of commitment is essential to convince farmers to transfer marginal land to forestry. This commitment would render well-founded arguments about long-term risk and replanting obligations redundant. The key therefore is continuity of State support especially in promoting the role of forestry in climate change. This requires incorporating forestry in carbon farming.

Because agriculture has been identified as the largest GHG emitter in Ireland, farmers feel under pressure to reduce production to achieve net zero by 2050, which is the aim of Ireland's *Climate Action Plan: Securing our Future*. Yet, there are major benefits for farmers who are prepared to combine forestry and agriculture.

Apart from the revenue from wood, farmers could maximise carbon production and income through afforestation according to a report by the Society of Irish Foresters. It proposes an afforestation scheme that will leverage the carbon value sequestered by forests and use this to reward and incentivise farmers to plant some of their farms.

A woodland carbon code provides evidence of the amount of CO_2 captured from the atmosphere, which varies depending on forest rotation, tree species and management regime. Selling the rights to the carbon captured would provide a long-term income for forest owners and remove the objection by some farmers to the replanting obligation, which commits their land to forestry in perpetuity.

A critical first step is to establish a forest or woodland carbon code for Ireland which is tailored to meet the needs and demographics of the Irish market and the farming community. The code would provide farmers with a carbon trading platform, whereby they could sell their carbon to the State or into the voluntary markets. The report puts a value on carbon, based on the Irish government shadow price of €32 per tonne in 2021. (A shadow price is an estimated price for something that is not normally priced or sold in the market.) This would provide a potential revenue of €7,910/ha for the carbon sequestered in addition to other income streams. The carbon value is a conservative estimate as the government's strategy is to increase the shadow price of carbon to €100/tonne by 2030 and €265/tonne by 2050.

Carbon values of this scale are more than sufficient to incentivise farmers to afforest areas of at least 10,000ha per annum. The impact on agricultural production is unlikely to be significant, based on previous experience where marginal areas of farms have been planted with a minimal reduction in the numbers of livestock units. The aim of the carbon afforestation scheme is not to displace agricultural production but to integrate carbon-based afforestation within the farm enterprise. In instances where farmers wish to maximise forest carbon – and timber – revenue, there would be a reduction of livestock numbers, which has the added benefit of reducing livestock emissions. The development of a forest carbon code would be a tangible acknowledgement of both the public good element of forestry and the owner's long-term commitment to an enterprise that has a legal replanting obligation in perpetuity.

Afforestation is a major commitment by farmers who traditionally manage crops for a seasonal return. Establishing short rotation forestry – up to 40 years – is a brave undertaking, even allowing for generous State support for the first 20 years for farmers and 15 years for non-farmer investors, but committing to a 100-year rotation oak forest demands a different and more sustainable partnership between the owner and the State where the ecological overrides the economic. This long-term collaboration is more akin to Edmund Burke's societal "contract", which "becomes a partnership not only between those who are living, but between those who are living, those who are dead and those who are to be born".

The EU Forestry Strategy has the support of the Commission in supporting carbon forestry by adopting "the carbon farming initiative, announced in the Farm to Fork Strategy, which aims to further promote a new green business model that rewards climate and environment-friendly practices by land managers, including forest managers and owners, based on the climate benefits they provide. Irish farmers as forest owners are best positioned to avail of State and EU supports in developing sustainable forestry to achieve the ambitious targets in Ireland's Climate Action Plan.

FORESTS AND LANDSCAPE

You must have something new in a landscape as well as something old, something that's dying and something that's being born.
Andy Goldsworthy

Because forest cover in Ireland has almost quadrupled between 1970 and 2020, emerging forests have become much more visible in the landscape even within a generation, especially where afforestation has been greatest.

Compared with Europe, forestry in Ireland, for all its benefits, is sometimes viewed as a controversial encroachment in the landscape. Conversely, in most European countries there has been little change in the ratio between forest land and agricultural land over the past half century. Also, the land use share in Europe between woodland and agriculture – grassland and crop farming – is unlikely to change, as there is little room to expand the forest area. Forest cover in EU Member States is 37.8% of the land area, with an additional 7.1% classified as shrubland (Figure 5.5). Ireland's land use differs from Europe in two major areas. We have high reliance on grassland (51.5% of the land area compared with 20.7% in the EU) while our forest cover is less than one-third of the European average (Figure 5.6).

In Ireland, there is a land bank of some 500,000ha of mainly marginal land available for forestry. Converting this to forestry would see grassland reducing from 51.5% to 44.5%, with a species mix comprising mainly high-yielding productive forests. While these are the kinds of forests most farmers, investors, timber processors and those employed in forestry need to provide revenue and much needed timber, they are not necessarily the forests the public want. An increase in forest cover to 17% of the land area as envisaged by successive governments will still result in a landscape with less than half the average tree cover of EU Member States. It still represents a significant change to a landscape dominated by grassland.

Comparisons between European and Irish forestry have limited merit, especially when issues such as landscape values are assessed. The landforms of an Irish landscape are different from other European countries where forest, woodland and shrubland are major landscape features. As a result, there are limited opportunities to increase forest cover in most European countries, unlike Ireland where a 60% increase in forest cover is a strategic government objective. The biggest challenge facing Irish foresters in landscape planning is how to integrate "new forests" with the natural and cultural landscape in a country which has fragmented semi-natural woodlands but no natural or primeval forests. While Europe too has lost almost all of its primeval or natural forests, its existing, mainly native forests bear a strong resemblance to the primeval even where man-made. Unlike Ireland, Europe is relatively rich in woodlands categorised as semi natural while Ireland has only 2.5% of land under native species including 1% of the land area

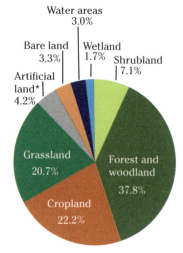

Figure 5.5: *Land use in EU Member State*

Source: Eurostat, 2018.

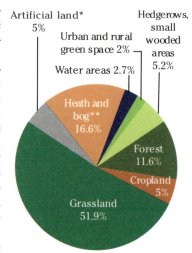

Figure 5.6: *Land use in Ireland*

Adapted by D Magner from Ireland's Forest Inventory 2022.
  *Built land
**Including cutaway peat.

covered by native semi-natural woodlands. Ireland's semi-natural woodlands bear little resemblance to the majestic primeval forest. They are managed mainly with conservation objectives to restore them to a state that resembles the primeval. In their present state, they are classed as components of the cultural landscape, which also includes Ireland's field system, nature reserves, forests and other topographical features. For the most part, these are not nearly as old as sometimes imagined. Our field system was shaped as early as the early Christian period but most of our fields and hedgerows were formed during the eighteenth and nineteenth centuries. The semi-natural woodlands around the lakes of Glendalough have tenuous links with the primeval, but eighteenth-century engravings show an almost treeless landscape. The cultural landscape, while man-made, has been created – or even recreated – over centuries. We have grown accustomed to it because it feels natural except when or our expanding villages, towns, cities, road network, new forests and other man-made features intrude.

How existing and new forests merge into this landscape is a challenge facing foresters and landowners, especially farmers, assuming they will carry out most of Ireland's afforestation programme. Forest owners have opportunities to change the design of their forests by opting for continuous-cover forestry on wind firm stable sites or by reducing the scale of final harvest by opting for small rather than large felling coupes. This approach provides opportunities for the introduction of mixed aged and mixed species during future forest rotations.

Scale is one aspect of forestry development that has greatest impact on the landscape. The large-scale State afforestation programmes carried out for much of the twentieth century in areas such as the Ballyhoura Mountains, Cloosh Valley, the Slieve Blooms, Slieve Aughty and Nephin combined social and economic forestry. Many of these, were established to address widespread unemployment and emigration in the last century. Despite difficult site and climate conditions, these forests fulfilled economic and social objective but large-scale planting of low nutrient exposed mountain and blanket bogs is no longer regarded as a sustainable land use and hasn't been practised since the end of the last century.

While many State forests established in the twentieth century are large by Irish standards, they are minuscule compared with forest tracts in Scandinavia, the Black Forest in southwest Germany and the man-made Landes Forest in France, which on its own is almost double the size of the total forest estate in Ireland. These immense European forests are integrated in their landscapes historically, culturally, spatially and topographically. They are accepted because nobody now living in heavily forested continental Europe can remember a time when forests did not exist and dominate the landscape. This contrasts with the Irish experience where the forest estate doubled in area between 1990 and 2021, from an admittedly low base, so the change in land use from agriculture to forestry has not gone unnoticed. Scale is the difference. Yet, scale is important in planning the Irish landscape, not just because of its uniqueness but also because of its intimacy and familiarity.

The single biggest change in Irish forestry has been the shift from State to private – mainly farmer – afforestation. This decision coincided with State and EU policies to encourage farm forestry, which unlike most European countries was virtually non-existent in Ireland but has replaced State afforestation since the 1990s. This has had landscape implications as afforestation is small-scale and is situated on better quality land than was made available to the State. Forestry now presents opportunities for farmers to integrate forestry and agriculture, not just as complementary revenue generation businesses but also as compatible land uses. It also allows forests to grow organically and to enhance the agricultural landscape.

Suitable land is available for forestry as outlined in reports by Niall Farrelly and Gerhardt Gallagher, and COFORD, Department of Agriculture, Food and the Marine (Table 5.2). They identify close to an additional 500,000ha of land suitable for forestry in Ireland. The bulk of the land identified, comprises 105,006ha of rough grassland and 179,000ha of unenclosed land without environmental restrictions. There are also opportunities for establishing small-scale mixed-species woodlands on better quality farm land – wet areas, field corners, etc, – as well as some upland and acid-designated areas where planting broadleaves and Scots pine could apply in programmes similar to the restoration of the Caledonian Forest in Scotland. In all, 490,000ha could be made available for afforestation without negatively affecting agricultural output.

The past afforestation of peatlands needs greater research to avoid forestry becoming a carbon emitter rather than a sequester, bearing in mind Ireland's *Climate Action Plan* findings: "Forestry on peat soils also generates emissions, and the extent of these lands in Ireland's forestry sector is significant, representing approximately 38% of the forest land category." Ideally, most of this land will be planted by farmers, which requires a different approach to tree species and landscape planning compared with earlier large-scale State forestry programmes. It needs to take account of existing land use, field systems and scale. Average farm size in Ireland is now 38ha while new forests are mainly established in relatively small blocks – averaging less than 8ha. While scale is the greatest difference between State and farm forestry, there are other differences, including location and the fertility of forestry land. Although most land selected for forestry is marginal to agriculture, farm forests provide greater options for tree species diversity and better opportunities for forest landscape planning and design that is in sympathy with the surrounding landscape. The *Forestry and the Landscape Guidelines* publication provides recommendations that acknowledge this potential, especially in relation to Ireland's landscape character types, including mountain, rolling moorland, undulating fertile farmland and drumlin.

Table 5.2. *Land use in Ireland and potential land availability for forestry (million ha).*

FORESTRY POTENTIAL KEY	No potential		Limited potential	Excellent potential	
CURRENT LAND USE (%)				POTENTIAL FOR FORESTRY	
1	2	3	4		5
EXISTING LAND USE (%)	DESCRIPTION OF LAND USE AND % OF TOTAL LAND AREA	AREA	COMMENTS ON FORESTRY POTENTIAL		AREA
Productive agriculture (34.2%)	Cattle (14.5%); dairy (8.6%); sheep (4.2%); tillage (5.2%); mixed livestock (1.7%)	2.39	Approx. 97% to be retained in agriculture apart from small pockets (3%) suitable for forestry including agroforestry.		0.07
Limited agriculture land (15.4%)	Cattle (7.5%); dairy (2.9%); sheep (4.2%); tillage (0.3%); mixed livestock (0.6%).	1.08	Revenue from forestry exceeds most sheep and cattle systems (Teagasc) – at least 20% of this land is suitable for forestry.		0.20
Poor agricultural and upland areas not farmed (4.1%)	Comprising poor grassland (1.5%) and unenclosed land (2.6%), both ideal for productive forestry.	0.29	Productive forestry land with few environmental constraints – 55% could be planted. No negative effect on agricultural production.		0.16
Land permanently unavailable (10.6%)	Land "biophysically" unavailable: water (2.5%); electricity and transport, (2.2%); urban and buildings (5.5%); scrub (0.4%)	0.74	Not available for forestry except small pockets of urban forests.		0
Land biologically and silviculturally unsuitable for forestry (12.2%)	Rocky land (3.3%); coastal sands and salt marsh (0.1%): raised bog/fens (1.5%); deep peat (4.3%); heathlands and other unproductive unenclosed land (3.0%).	0.85	While some tree planting would benefit parcels of this land, they are best retained in current state apart from a small percent of cutaway bogs.		0
Severely restricted forestry land due to National and EU designations (12.9)	Designated hen harrier (0.8%); Natura 2000, NHA and Nature Reserves (2.6%); Fresh water pearl mussel (all catchments) (7.3%); potentially acid sensitive (2.2%)	0.90	Most of this land to be retained as Natura 2000 and acid sensitive sites but 5-10% native species could be planted in some pearl mussel catchments and acid sensitive areas.		0.06
TOTAL UNPLANTED (89%)	**Existing unplanted land**	6.25	**Potential forest area (7%)**		**0.49**
TOTAL FOREST (11%)	**Existing forests and woodlands**	0.77	**Existing forest area 2021 (11%)**		**0.77***
TOTAL ALL LAND (100%)	**Total land area – forest and non-forest**	7.02	**Potential forest cover this century. (18%)**		**1.26***

Source: Adapted by D. Magner from *Land Availability for Afforestation*, COFORD (2016) and "An analysis of the potential availability of land for afforestation in the Republic of Ireland" Niall Farrelly and Gerhardt Gallagher (2015).
*This area does not include existing 347,690ha of hedgerows and small scrub/wooded areas (4.9%).
Note: This study was based on a total forest area in of 0.77 million ha in 2021. This forest area increased to 0.80 million ha according to the *2022 National Forest Inventory.*

FORESTS AND BIODIVERSITY

Biodiversity – or biological diversity – is a relatively new term to describe living organisms and their habitats. In 1992, it was defined by the Convention on Biological Diversity as:

> … the variability among living organisms from all sources including, *inter alia*, terrestrial, marine and other aquatic ecosystems and the ecological complexes of which they are part; this includes diversity within species, between species and of ecosystems.

Forest ecosystems are major biodiversity habitats not only for tree species but also for a diverse range of flora and fauna, many of which are rare or threatened. Globally, forests are home to half of all terrestrial species, according to the World Wildlife Fund (WWF). Forest canopies are among the most species rich terrestrial habitats supporting about 40% of invertebrate species.

Conserving biodiversity in natural forests is a key element in the protection and enhancement of their unique ecosystems. But natural forests, with their untouched or pristine ecosystems, are rare globally and no longer exist in Ireland, even in our protected nature reserves, national parks, bogs, bare mountains and the few remaining semi-natural woodlands. Human and climatic intervention have changed the ecosystem beyond recognition since the first hunter-gatherers arrived in Ireland some 12,500 years ago.

This is especially true of the primeval woodland – including its associated ecosystem – which has vanished without trace. Even the fragmented semi-natural woodlands that remain, bear little resemblance to the primeval but are worth protecting, enhancing and expanding. Leaving these semi-natural ecosystems alone to return to nature or to get on with the business of reconnecting with their once-natural habitat is not an option. In woodland conservation projects, including rewilding, enlightened human intervention is vital in recreating native woodlands. Otherwise, introduced tree species outcompete and outnumber natives.

In addition to trees, forest flora includes vascular plants, bryophyte and lichen floras, mosses and fungi. The variety of forest flora depends on a number of factors including the underlying soil, tree species, tree spacing and forest age.

The major difference between Irish and European forests is the relative paucity of native species diversity, both flora and fauna. Yet, Ireland is host to a vast range of vascular plants – including trees species – but few are native. Likewise, over 450 bird species are recorded in Ireland but Ireland has considerably fewer breeding birds than European countries, including Britain.

How native species are protected and enhanced to ensure compatibility alongside introduced species is a major environmental and land use challenge. It exists in forestry where

BIODIVERSITY AND BIOCLASS

Private afforestation is now expected to allocate 35% of the forested area to native species and open unplanted areas or Areas of Biodiversity Enhancement (ABEs). These are largely unproductive but ecologically rich so they require a different management approach than productive forests.

In 2018 Coillte designated 90,000ha of its forest estate as biodiverse habitat. This is a considerable area when compared with 60,000ha of protected raised and blanket bogs in Ireland. The company has developed a BioClass system which provides a range of information on tree (forest, woodland and scrub) and non-tree (bog and heath) habitats. The BioClass area of the Coillte estate is divided almost equally between forest and open habitats: forest habitat comprises native forest (11%), broadleaves (7%), mixed forest (8%), conifers (15%) and scrub (7%), while open habitat comprises bog (28%), heath (23%) and other habitat (1%).

The allocation of 20% of the Coillte estate as biodiverse habitat prompts the question about the ecological potential of the total forest estate in Ireland. The 2017 National Forestry Inventory (NFI) estimated that the biodiversity resource in the total forest estate comprises 35% broadleaves and open area including "a high prevalence of non-tree plant species and lichens" while "nearly half (49.3%) of the forest area has vegetation coverage of greater than 90%".

introduced tree species far outnumber our limited range of native species. This dearth of native species is amplified in agriculture where there is total dependence on introduced species in crop production and animal husbandry.

It is important therefore to acknowledge the limited palette of native flora and fauna in Ireland. While a case has been made to reintroduce extinct mammal species, the best option is to work with what we have. Reintroduction doesn't arise in forestry, as all tree species that made it to Ireland after the last Ice Age still exist in our landscape, even though some may be threatened by disease.

By creating the right forest and woodland environment, we can protect native flora and enhance habitats for native bird life. For example, we have seen a return of a number of bird of prey species to Ireland's forests and woodlands; some of which were absent for over a century. It is important therefore to assess the forest as an ecosystem that contains not only trees but flora – both vascular and nonvascular plants – fauna, including forest mammals and birds as well as invertebrates such as butterflies, spiders and other insects. In doing so we need to enhance our limited range of native species and acknowledge our exciting range of introduced species so that a biodiverse equilibrium is achieved and maintained.

Tree and flora diversity

The contrast between the number of native vascular species growing in Ireland compared with Europe is enormous. For example, there are an estimated 11,000 vascular plants which are native to Europe but only a small fraction of these made it to Ireland before the land bridge connecting Ireland to Britain was submerged towards the end of the last Ice Age. Estimates on the numbers of native and naturalised vascular plant species vary. *Ireland: Red List No. 10: Vascular Plants* considers "a total of 1,211 taxa, comprising 1,047 species, 4 species aggregates, 157 subspecies and 3 interspecific hybrids" as "native, archaeophyte or of uncertain native/alien status". David A. Webb maintains that only 815 vascular plants are native. T.G.F. Curtis and H.N. McGough believe this to be a conservative estimate, maintaining "the total may be as high as 1,000". Considering Ireland's size, we have a rich variety of indigenous and introduced plant species and subspecies, now estimated at 1,543 by Irish botanists John Parnell and Tom Curtis. Add these to nonvascular native plants, and the diversity is considerable, as explained by Jenny Neff: "... the number of taxa in Ireland for any given group of plants is low in comparison to our neighbours, but in some groups, it is clearly rich, e.g., bryophytes and stoneworts".

The list of Ireland's native tree species is even more limited than other vascular plants. We have only 20 native tree species if the accepted definition of a tree is a woody plant, capable of growing to at least 6m in height and usually – but not always – with a single stem or trunk. This is well below Britain's 35

Native birch woodland in Caragh Lake Forest, Co. Kerry, one of the biodiverse habitats identified in Coillte BioClass forests.

native species – identified by John White, Jill White and S. Max Walters. Both islands compare unfavourably with continental Europe where 454 native species are catalogued in the European Red List.

In Ireland, you could count on one hand the number of native tree species that provide viable productive woodland. Oak and Scots pine on suitable soils and cherry in mixture all have proven characteristics, while alder and birch adapt well to a wide range of soils but have limited usage. Sadly, that's it, now that ash and elm are no longer planted due to disease vulnerability. The remainder have undoubted ecological potential, but commercial timber production is not their forte.

Although the main species in this section of Dromkeen Wood, Co. Cork are non-native Japanese larch, Douglas fir, beech and Sitka spruce, they are now accepted as an integral part of the Irish landscape and like other introduced trees they grow exceptionally well in Ireland. The understorey comprises a mix of native plant and tree species including woodrush, fern and holly.

All European indigenous spruce, larch, fir and pine species (with the exception of Scots pine) are non-native to Ireland. Broadleaves, including beech, sycamore, sweet chestnut, field maple, Norway maple and lime also failed to colonise Ireland naturally and were introduced by humans over the centuries.

Fortunately, all these European species grow well in Ireland, while many European broadleaves are now naturalised. Even better performances are being achieved by coniferous species that originated in western North America, especially where climatic – maritime – conditions are not dissimilar to Ireland (see Chapter 8). The challenge when introducing European and Pacific Northwest species is to ensure that they not only provide productive forests but that they are managed in a way that enhances, and not overwhelms, the ecosystem and landscape. While some of the European exotics may have been introduced by the Normans, the American species are later arrivals. These were introduced in the early nineteenth century and only planted as forest trees a century later. While the economic worth of these coniferous plantations is acknowledged, their biodiversity role has been criticised but may well be underestimated as Sandra Irwin states in the study. "The value of plantation forests for plant, invertebrate and bird diversity and the potential for cross-taxon surrogacy":

> The species richness of non-native spruce-dominated plantations can be as high as that found in semi-natural woodlands, which suggests that temperate plantation forests, with appropriate management, can provide habitat for plant and animal species.

The term "appropriate management" used by Irwin is important. Forest owners, foresters and landscape planners also need to heed the study's findings that "oak and ash woodlands support different communities than plantations, which must be considered in forest management for environmental objectives". The introduction of more environmentally positive management regimes and greater emphasis on small-scale mixed species afforestation is in tune with the author's hypothesis:

> Our findings demonstrate that careful management of non-native conifer plantations is required to enhance species diversity and optimise their contribution to landscape scale biodiversity while preventing negative biodiversity impacts.

Because of Ireland's low forest cover, there is ample space to increase multi-species and mixed-aged forests that allow wildlife corridors – including hedgerows – and open spaces to enhance biodiversity. Biodiversity is also improved by good forest management practice, which requires looking objectively at past successes and failures. It is accepted by foresters and environmentalists alike that large tracts of single-age, single-species coniferous forests reduce biodiversity. For a period in Irish forestry, during the twentieth century, a number of large-scale forests were established using single species but this is no longer the case, as all forests must now carry a mix of at least 35% broadleaves and open spaces while the scale of forestry development has reduced considerably as the average new plantation size is approximately 8ha.

Table 5.3: *Composition of the total forest area (1,000s ha) and per cent forest cover by species and bio-diverse unplanted areas.*

Forest composition	Area	%
Sitka spruce	360.9	44.6
Norway spruce	27.0	3.3
Scots pine	8.4	1.0
Other pine species (mainly lodgepole pine)	62.8	7.8
Douglas fir	9.3	1.1
Larch species (mainly European and Japanese larch)	23.8	2.9
Other conifers	2.9	0.4
Total conifers	**495.1**	**61.2**
Pedunculate and sessile oak	20.2	2.5
Beech	10.7	1.3
Ash	24.3	3.0
Sycamore	10.5	1.3
Silver and downy birch	58.0	7.2
Alder species	19.7	2.4
Other short-living and understorey broadleaves (mainly willow, hazel and mountain ash)	63.3	7.9
Other long-living broadleaves (including holly, sweet chestnut, lime (large–and small–leaved) and cherry)	11.4	1.4
Total broadleaves	**218.1**	**27.0**
Total tree cover	713.2	88.2
Forest open areas (reparian zones, roads, trails, etc.)	88.1	10.9
Temporarily unstocked – awaiting planting	7.6	0.9
Total forest area	**808.9**	**100**

KEY TO SPECIES' ORIGINS

Native

European and naturalised

Exotics – mainly western North America

Mixed origins

Source: Adapted by D. Magner from Ireland's National Forest Inventory 2022,

Even allowing for predominantly coniferous planting, the forest estate still maintains a higher species range than it is given credit for. There is a misconception that the Irish forest estate is "wall-to-wall" Sitka spruce. True, Sitka spruce is the dominant species but it constitutes 44.6% of the forest estate or just 4.9% of the area of Ireland (Table 5.3). The remainder of the forest area includes mainly native broadleaves (27.0%), diverse conifers (16.6%), and open areas (10.9%) such as roads, pathways, hedgerows, riparian zones and firelines.

This species mix allows foresters and forest owners flexibility to maximise biodiversity while ensuring productive, commercially viable forests. It is important in achieving these two objectives that the benefits of commercial coniferous forests are acknowledged in the species mix. These benefits are considerable, if managed sustainably.

MAAM VALLEY ANCIENT WOODLAND RESTORED

It is difficult to imagine that Maam Valley – or Joyce Country – was once covered with trees. Now, apart from the coniferous forest blocks to the west of the road that skirts the meandering Joyce – or Maam – River, the valley is almost treeless. At first glance, there are no trees to the east of the road, but on closer inspection there is a native woodland in the making at Griggins townland. Established in 2018, this is the brainchild of Marina Conway, Western Forestry Co-operative who along with visionary farmer Micheál Laffey (above) is reimagining the primeval forest that once dominated this landscape. Laffey converted 15ha of his hill farm to native woodland to improve and protect the soil, enhance the landscape, increase biodiversity and provide timber in the future.

Birch and alder provide soil improvement and nurse benefits that benefit oak. Birch leaves help develop mull humus which raises soil pH while alder has nitrogen fixing qualities which allows it to take up atmospheric nitrogen. Other species such as Scots pine, mountain ash and willow fit seamlessly into this rejuvenated landscape, which begs the question "why aren't more suitable upland sites and unenclosed land afforested with native species?" as advocated by Western Forestry Co-op, IFA and the Society of Irish Foresters.

Creating the right balance between the economic and the ecological is achievable by acknowledging both the benefits and the risks of productive plantation forests. In their 2013 study "Do Irish Forests Provide Habitat for Species of Conservation Concern?" Sandra Irwin and fellow authors, concluded:

> This intensive survey of plantation forests throughout Ireland has revealed that these sites offer habitats to some nationally rare and threatened species of plants, invertebrates and birds. The potential of plantation forest to enhance national biodiversity is thus evident and so the planning and management of these forests should incorporate this goal.

The authors, conscious of the inherited dearth of native woods in Ireland added the proviso:

> Our results also indicate that biodiversity conservation measures should target the expansion and restoration of semi-natural woodlands (including conifer removal); these sites may be important refuges for forest-associated species in intensively managed landscapes.

Thus, while they acknowledge the role of plantation forests in the provision of habitats for biodiversity, they place strong emphasis on conservation. They call for "the retention or restoration of [semi-natural woodland] habitats for forest biodiversity". The innovative Native Woodland Schemes for afforestation and conservation, coupled with afforestation programmes which set a minimum of 35% broadleaves and open space should fulfil these criteria at a national level.

Forests have an added advantage in encouraging biodiversity as good forestry practices either minimise or completely avoid artificial growth stimulants or chemical control, while fungicides and insecticides are never applied in establishing new forests. Native woodland schemes (see panel on Maam) are completely chemical and fertiliser-free. When fertilisers and herbicides are required, these are applied only once in forest rotations which can extend from 30 to over 100 years. It is planned to phase out even the once-off selective use of chemicals to prevent pine weevil damage in reforested sites after final harvest.

Unlike intensive agriculture and other land uses, virtually all forests are established successfully without nitrate or potassium applications, while phosphate in the form of one application of rock phosphate can be applied during planting. This is a slow-release fertiliser, which benefits young trees over a number of years. Essentially, all that trees and forests require are sunlight, rainfall and soils of even moderate fertility.

The common perception, that coniferous forests have less understorey plant variety than broadleaved woodlands is largely true but varies with species, age and management practices. With a few exceptions, understorey plant variety increases as forests grow older, regardless of tree species. Also, ground vegetation cover varies widely between and within individual species – this is true for both conifers and broad-leaves. For example, conifers such as Scots pine and larch have a rich variety of understorey species compared with all broad-leaves and much more than beech, which tolerates little ground vegetation competition. On the other hand, native and natural-ised species including oak, ash, birch, alder, sycamore and sweet chestnut provide much greater diversity of plants and understorey shrubs than most conifers.

Open areas such as setback from walls, in the Devil's Glen Wood (op-posite) or other features are left un-planted to allow the forest to create its own ground vegetation layer such as bluebells which colonise this ring fort site (below) in Curragh wood-land, Co. Cork.

Conifers provide greater shade than broadleaves so the ground vegetation layer has reduced sunlight for vigorous growth of plants and shrubs, especially when the crop closes canopy. Ground vegetation is encouraged as the crop matures and regular forest thinning reduces tree density and allows sun-light to reach the forest floor.

Reduced flora also depends on soil type and forest loca-tion. Flora diversity decreases as soils become more acidic and nutrient-poorer. These are generally the soils where most of the coniferous forests in Ireland were located during the twentieth century. However, with careful management and judicious tree

FOREST FUNGI

The relationship between fungi and trees is complex as they can be both beneficial and destructive. Fungi can destroy trees and forests as witnessed by the pathogen *Phytophthora ramorum*, which kills larch and other species while the *Hymenoscyphusfraxineus* or ash dieback is wiping out Irish and European ash species. Native and naturalised elm have almost disappeared from the landscape as a result of Dutch elm disease, caused by two related fungi: *Ophiostoma novo-ulmi* and *Ophiostoma ulmi*.

But there are major positives to forest fungi. They assist in the breakdown of material in dead trees and play a vital part in forest regeneration. There is the unseen fungus – mycorrhiza – that helps trees take up nutrients more efficiently through their root system than trees where this fungus is absent.

The visible fungi or mushrooms are a rich source of food, providing the forager knows the difference between edible, inedible and poisonous varieties. As our forest estate expands and matures, fungi variation will increase as litter fall and decayed wood on the forest floor provide excellent fungal habitats.

Overleaf: The attractive but poisonous fly agaric fungus in a predominantly Norway spruce forest, with some beech and other broadleaves retained in Clonad, Co. Offaly. The rich litter layer and ample deadwood provides an excellent environment for a wide range of fungal species. See also page 224.

species selection, soil amelioration is possible by introducing species such as birch, which will over time encourage greater flora diversity.

While the range of native vascular flora is limited in Ireland, John Cross and Kevin Collins outline a species richness in "the bryophyte and lichen floras":

> … Ireland has one of the richest moss and liverwort floras in Europe and some woodlands especially in the west of the country, contain more species of these diminutive plants than flowering plants.

The authors are referring to flora in native woodlands but well-managed forests featuring naturalised and exotic tree species provide rich habitats for these species, especially mosses and lichens, as well as distinctive forest fungi.

Leaving open areas is important to protect flora diversity, especially in plantation forests. While these unplanted areas are usually created to enhance bird and mammal diversity, they are also important in maintaining and improving the range and variety of flora.

Forests established on enclosed land can maintain and improve plant diversity by leaving unplanted land beside hedgerows and older mature trees as well as open areas such as setback along boundaries, rivers and streams. These act as biodiversity corridors and are an intrinsic element of good forest design. Plant diversity is not threatened where former grassland sites are planted because plant diversity is limited on these

sites to begin with, especially after periods of intense farming. However, even where industrial agriculture has been practised, hedgerows have remained rich in flora and fauna diversity because they haven't been subject to fertiliser, herbicide or insecticide applications. Setback areas along hedgerows should be a feature of forest layout. Because planting sites are relatively small in Ireland – averaging just under 8ha – there are opportunities to create a mosaic of land types linking rivers, plantation forests, agricultural land and old woodland where trees and vascular plants are nurtured alongside grasses, flowering and other plants.

Forest and fauna diversity

Deforestation, agricultural development and bog formation since the arrival of humans to Ireland have changed the terrestrial landscape to such a degree that none of the primeval fauna habitats survive.

Centuries of agricultural development, especially mechanised farming, forest clearances and other habitat loss have radically reduced Ireland's native terrestrial fauna. In addition to human intervention, climate change may also have caused the extinction of animals such as the giant Irish deer, reindeer and lemmings. Land use change is likely to have caused extinctions of woodlark, white-tailed eagle and golden eagle by the early twentieth century.

Like its restricted range of native vascular plants, Ireland also has a limited number of indigenous mammals and birds. As a result, issues such as the reintroduction of once-native bird and mammal species are constantly debated. A number of successful reintroductions of raptors have taken place in Ireland, but calls to reintroduce extinct animals such as the wolf have – on balance – been wisely resisted.

While the debate on reintroducing mammals and bird life is likely to continue, protecting existing native and naturalised species ought to be a priority. Achieving this objective in an ever-changing forest landscape is a major challenge for forest owners and foresters.

Citing a number of studies, the Ireland's Red List (No. 12) states: "Of the 219 terrestrial mammals found in Europe, only 27 or 12.3% are native or long established in Ireland, compared with 43 species found on the island of Britain." Ireland's forests provide habitats for most of our terrestrial mammals as discussed by Seán Rooney and Thomas J. Hayden:

> Ireland's broadleaved and coniferous forests provide habitats for a range of mammal species, from large and medium sized herbivores such as red, fallow and sika deer, to smaller mammals such as pine martens, red and grey squirrels, bank voles, wood mice and shrews. In addition, hares and rabbits are frequently found at the edges of such habitats.

The role of native and naturalised woodlands versus mainly exotic coniferous forests in providing biodiverse

REINTRODUCING BIRDS OF PREY IN IRELAND

Species reintroduction although rare, is not new to Ireland. The red squirrel was reintroduced in the early 1800s after its extinction, probably in the 17th century. Recent debates on bird and mammal reintroductions have proved controversial. For example, suggestions to reintroduce mammals such as the lynx and wolf have been strongly resisted while there is a more benign attitude to reintroducing birds. A number of successful reintroductions of birds of prey have taken place in Ireland, including the golden eagle, white-tailed eagle, red kite and, most recently, the osprey (below). In July 2023, 12 osprey chicks were introduced from Norway and released in southern Ireland by the National Parks and Wildlife Service (NPWS). The aim is to establish a viable, free-ranging osprey population that will eventually breed in Ireland, 150 years after its extinction. The NPWS plans to add up to 70 osprey chicks from Norway over a five-year period.

Ornithologist Gerry Murphy says that well managed forests are excellent habitats for ospreys as experienced in Europe: "Ospreys predominantly depend on tall trees – conifers or broadleaves – to build their nests. Some nests, in their current breeding ranges in the continent, have been in use for approximately 20 years, with the birds adding to it each year."

Since its introduction to Ireland in 1911, the grey squirrel has spread throughout most counties and has threatened to wipe out the red squirrel population.

habitats that protect and enhance fauna is misunderstood in Ireland as is the dynamic between forest mammal species, which changes as forests evolve. This is best illustrated by the fluctuating populations of native red squirrel, pine marten and the spread of the introduced destructive grey squirrel.

It was feared at the turn of this century that the red squirrel would be wiped out by the grey squirrel, which had increased in numbers and range since its introduction to Ireland in 1911. There were good reasons for this alarm as grey squirrels compete with red squirrels for food and breeding habitats and also carry a disease – Parapox virus – that is fatal to red squirrels. In 2007, Michael Carey reported "some spread" of red squirrel populations in parts of counties Offaly, Kildare, Cork and Donegal, but warned of its rapid decline:

> Unfortunately, it may now be considered extinct in Meath and West meath, and has become particularly rare in Kilkenny, Carlow and Louth. Red squirrels in areas where greys are already established are under particular threat, while other populations just beyond the current grey squirrel distribution (such as Cork, Limerick, Kerry and [northeast] Antrim, where significant areas of mature broadleaved woodland exist) may also be considered at risk. The speed of grey spread suggests that it could colonise these areas in 10-20 years.

Pine marten was also considered a rarity by then. Yet, within 12 years there was an increase in populations of both pine marten and red squirrel as reported by Colin Lawton and fellow authors:

> Pine marten continues to return following its previous decline across the island, and is now found in every county, having been recorded on six occasions in Co. Derry, where it was considered absent in 2012. The core range of the pine marten, which has been correlated to the demise of the grey squirrel, has expanded and now stretches through the west, the midlands, the southeast of Ireland and parts of Northern Ireland. This suggests that the grey squirrel may start to further decline in these areas in the coming years.

The expansion of coniferous forests has been a key element in the increase in pine marten populations, which has led to a decrease in its grey squirrel quarry and a resultant increase in red squirrel distribution. Who knows what the next phase of this symbiosis may bring because nature can be unpredictable and cruel. Natural history tells us the pine marten was present in Ireland when the red squirrel became extinct, probably in the beginning of the seventeenth century before it was reintroduced in the 1860s. Once the pine marten has dealt with the destructive grey squirrel, could it turn its attention to the native red? Gerry Murphy believes this is unlikely as the red squirrel is much lighter than the grey. The more nimble red squirrel can retreat rapidly to the tips of trees and escape the pine marten unlike the heavier grey. But the welcome decline in grey squirrel numbers may not be fully due to the pine marten. Irish ornithologists believe that the

The feared extinction of the native red squirrel by the introduced grey has not materialised. This is partly due to the increased numbers of its main predator, the pine marten.

reappearance, in ever-increasing numbers, of native buzzard populations may have helped the survival of the red squirrel. While these birds of prey can be observed soaring and gliding over open country, seeking rodents as food sources, they also feed on small mammals in forests and woodlands. In this environment, grey squirrels – which also forage for ground food, unlike red squirrels – can be easy prey for vigilant buzzards.

Both forest inscapes and forest edges provide rich habitats for native and introduced mammals, such as badgers, foxes, hares and rabbits while the larger herbivores – red, fallow and sika deer – have a more contentious relationship with trees and forests. Sika, in particular, causes serious damage by browsing small saplings and debarking young and semi-mature trees. The increasingly fragmented nature of Ireland's private forests, while providing diverse wildlife corridors, has also "facilitated the growth of deer populations to such an extent that they are now limiting the range of tree species that can be used in

The pine marten is native to Ireland and is making a natural recovery, especially in conifer forests such as Sitka spruce (above). A positive side effect of increased pine marten populations is a reduction in the numbers of introduced and destructive grey squirrel, which is followed by an increase in the native red squirrel population.

afforestation," according to Seán Rooney and Thomas J. Hayden. The challenge now is to balance habitat diversity and biodiversity enhancement with control of the main mammal species that damage trees, especially native broadleaves.

The value of plantations for groups such as beetles, hoverflies, spiders, moths and butterflies has been substantiated by studies in planted forests. For example, Smith and fellow authors acknowledge that "plantation forests do have the potential to cause some increase in hoverfly biodiversity at the landscape scale when they are located in intensively farmed landscapes". They acknowledge the most important habitats for hoverflies "are features associated with wet substrates and dead wood". Regarding forest species, they state:

> We found little overall difference in the hoverfly biodiversity of ash and Sitka spruce plantation forests, but ash forests do appear to support a greater number of saproxylic species. Adding ash to a Sitka spruce plantation is likely to increase the hoverfly biodiversity at the plantation scale, especially if the ash component includes grassy clearings.

While ash is no longer planted due to disease risk, most hoverfly and other insect habitats can be enhanced by good forest management practices such as mixing species, leaving dead wood and increasing canopy openings. Management practices can also influence bat habitats, as outlined by Tony Mitchell-Jones and adapted by Orla Fahy and Ferdia Marnell for Ireland:

> Forests of all types, ranging from the semi-natural forests to broad leaf and conifer plantations, and all ages from newly established to closed canopy forests, are used by bats. In many cases, bats will seek out particular features, such as ponds or streams, hedgerows, clearings, aquatic buffer zones, archaeological exclusion zones or forest edges, where insects tend to be most abundant.

Regarding bat species diversity, the authors say:

> Although forests/woodlands are used in some way by all ten Irish bat species, three species are considered to be woodland specialists, namely whiskered, Natterer's and Brandt's. Bat distribution, diversity and density in managed forests may be affected by competition for limited roosting places. Managing forests to maintain or improve bat populations needs a good understanding of the bat species present and their roosting, hunting and commuting needs.

The Irish Rare Birds Committee recorded up to 475 bird species at least once in Ireland or its marine waters since 1950. Bird-Watch Ireland identifies 450 species with some 200 'regularly occurring' bird species in Ireland. The 2021 review of bird status in Ireland by Gillian Gilbert, Andrew Stanbury and Lesley Lewis found that 54 of Ireland's regularly occurring bird species [211] are now on the Red – endangered – List.

Irish forests and woodlands are home to a wide range of birds, including nationally important populations of some rare, threatened or declining species including wood warbler, red-start, red kite and buzzard, while nightjar have been reduced to a few pairs. Ireland was the only country in Europe believed to have no resident breeding woodpecker, until 2006 when the great spotted woodpecker was observed along the east coast. Today, woodpeckers are once again breeding in our coniferous and broadleaved woodlands and expanding their range in ever increasing numbers.

Opposite: Long-eared owls in a Sitka spruce forest. They gain new habitats as coniferous forests mature especially in open areas within the forest or located close to fields and fallow land where they hunt for small mammals, birds and insects.

The reappearance, in ever-increasing numbers, of buzzard populations in Irish forests, may – in tandem with the pine marten – help in the survival of the native red squirrel.

The common buzzard is also going through a recolonisation phase after a significant absence from our forests. This tree-nesting bird of prey can now be found breeding in woodlands in almost every county of Ireland and with their wide variety of food choice will help keep rabbit, magpie, corvid and – as already mentioned – grey squirrel numbers in check.

An increasing number of forest specialists, such as cross-bills and siskin, are breeding in Ireland. This is due in large part to increased afforestation in recent times, maintains Donal

Whelan of the Irish Timber Growers Association and fellow authors in "Irish Forests and Biodiversity". In addition, sustainably managed forests provide habitats for bird species of scrub and open habitat, including those of conservation concern such as the grasshopper warbler, whinchat and linnet. In the absence of plantation forests, such species would be scarcer in the largely agriculture-dominated landscape of Ireland.

The long-eared owl is found in coniferous forests throughout Ireland, apart from some areas in the west. The species favours coniferous blocks of forests, preferably those located close to open grassland and fallow fields, where it hunts for a variety of prey including field mice, small birds and insects.

On the other hand, BirdWatch Ireland points to large scale afforestation as contributing to the decline of bird habitats, in particular the curlew and lapwing. Ireland's forestry programme (2023-27) contains safeguard and procedures that will protect these and other wader species. However, bird population numbers in forests vary widely. As 61.2% of forests comprise conifers, these have been the focus of attention of a number of studies such as the "Biodiversity of plantation forests in Ireland: BIOFOREST project" by John O'Halloran and fellow researchers carried out between 2001 and 2005. It concentrated on mainly State-owned coniferous – Sitka spruce, dominated – forests. It found that "more than 55 bird species were recorded during 2001 fieldwork". The authors made the following observations:

> Preliminary analyses indicate clear differences between the bird communities of young plantations (in which willow warblers and redpolls are abundant) and older forests (in which goldcrests and coal tits are more numerous). Some of the variation in the data appears to be related to tree species composition as well as to age. The two bird species recorded most frequently were wren and chaffinch. Wrens were recorded twice as frequently as chaffinches in all ages and types of forest except for mature Sitka, where approximately twice as many chaffinches as wrens were recorded, and mid-rotation Sitka, where numbers of chaffinches and wrens were approximately equal. Species richness and total numbers of birds also appear to vary with forest type and age.

The interaction between forests and changing bird populations is dependent on a number of factors including tree species composition, management practices, forest size and age structure. Ensuring optimum birdlife habitats is an ongoing challenge that will require changing management practices.

In their 2016 study "The role of planted forests in the provision of habitat: an Irish perspective", O'Callaghan and fellow researchers acknowledge the role of plantation forests as bird habitats, but they also point to deficiencies:

> We find that plantation forests provide a suitable surrogate habitat primarily for generalist species, as well as providing habitat for certain species of conservation concern. However, we find that plantation forests provide poor habitat for native forest specialists,

NEW BIRD ARRIVALS

The great spotted woodpecker (above), initially observed along the east coast from Co. Down to Co. Wexford in 2005, is now found in most counties. Like the arrival of great spotted woodpecker, there are other bird species that have arrived in recent times and are now nesting in broadleaves and conifers. For example, in recent times, the little egret was first recorded breeding in riparian woodlands at the lower reaches of the Munster Blackwater.

Species showing a good likelihood of settling here soon include, cattle egret, and great white egret (below), according to Gerry Murphy. Both species nest colonially in trees, bushes and reed beds, close to large lakes and rivers. Woodlands and forests adjacent to river systems such as the Munster Blackwater, Suir, Nore and Barrow, while large water bodies, such as the Midland lakes are likely to be colonised at the outset.

WADER PROTECTION

All afforestation projects in Ireland require assessment of their impact on the breeding waders of conservation concern. Afforestation applications require ornithological reports to demonstrate that planting a site does not result in bird habitat deterioration for the following waders: curlew (above), dunlin (below), lapwing, red shank, snipe and golden plover.

Assessment of afforestation applications require answering questions such as "is the planting site within the BirdWatch Ireland Breeding Water Hotspot map?" to ascertain if the project area contains suitable breeding and foraging habitat for the relevant breeding waders. Various conditions apply in wader protection. For example, planting is not allowed within a distance of up to 1.5km from a curlew nesting site. Waders of most significance in forest sites are curlew, lapwing, snipe and golden plover whose breeding range is north of a line connecting Limerick and Dublin cities with the exception of the area around the Kerry northeast and Cork northwest border.

and [we] examine potential management strategies which may be employed to improve habitat provision services for this group.

Many studies of forest habitats carried out in the late 1990s and the first decade of the twenty-first century concentrated on specific sites, mainly large blocks of State forests. Most of these were established from the mid-twentieth century until large-scale State afforestation was phased out by 1996 and replaced by private – mainly farmer – planting.

By 2017, the total forest estate was divided almost equally between State and private. The increase in private afforestation which began in earnest in the mid-1980s – overtaking the State in 1989 – resulted in fundamental changes to forestry development in relation to tree species selection but especially to location and scale. The average size of private forests established from 1980 to 2020 was 8.6ha – it has decreased to 7.5ha since then – which is minuscule compared with the large forests established by the State during the twentieth century. Afforestation of blanket peats was discontinued in the 1990s. In addition, the availability of better-quality land resulted in a change of tree species composition. For example, broadleaf planting increased from 5% of afforestation programmes in the twentieth century to 30% during the first two decades of this century. In this scenario, there are ample opportunities to further improve plantation forest biodiversity as proposed by O'Callaghan and fellow researchers on low-biodiversity habitats, including "retaining any complementary habitats such as hedgerows".

So, how does the emerging forest estate ensure that commercial forestry can accommodate the protection and enhancement of bird habitats? M. W. Wilson and co-authors provide some recommendations in their study "Bird diversity of afforestation habitats in Ireland: current trends and likely impacts":

> The bird communities of many shrub-poor and over-grazed grasslands, as well as some degraded peatland areas, are sufficiently impoverished that their replacement with the bird assemblages supported over the course of a forest cycle may, in many cases, constitute an increase in the avian diversity of these areas. This is especially likely to be the case if the management of these plantations takes birds into account. In particular, sufficient open space should be left around existing areas of shrub and tree cover to enable their persistence during later stages of the forest cycle. Increasing the rate at which agriculturally improved grasslands are afforested will support the government's strategic aim to increase forest cover, improve the ability of intensively farmed areas to act as carbon sinks and relieve the pressure to plant on more biodiverse habitats.

Private afforestation since the 1990s has dramatically altered the forested landscape resulting in a mosaic of uneven-aged small-scale forests and their relationship with the landscape. This development has potential to accommodate bird habitats including the hen harrier and merlin.

Forestry has been linked with the fluctuating fortunes of a number of bird species, including the hen harrier and merlin. The hen harrier had gone into decline in the second half of the nineteenth century and had almost become extinct by 1950. In 1981, Gordon D'Arcy, maintained in *The Guide to the Birds of Ireland*: "Were it not for the re-afforestation of this country in the [twentieth] century, this harrier would probably not be found here any longer." He went on to say: "The man-made forests now provide vital sanctuary for this beautiful bird of prey".

As a result, it was widely believed that the relationship between plantation forests and the hen harrier was positive. Many forest owners, foresters and other stakeholders were genuinely surprised that an area of 167,100ha was designated by the National Parks and Wildlife Service (NPWS) as Special Protection Areas (SPAs) under the Birds Directive on account of its importance for holding hen harrier breeding populations, as habitats were in decline. While hen harriers were breeding in young plantations, they "were not to be found in 'worked' or mature plantations", as D'Arcy explained in 2014.

While the IFA accepts the designation of hen harrier SPAs, Michael Fleming, IFA forestry committee chairperson, farmer, and forestry contractor believes that 127,000ha would represent a suitable land bank for hen harrier conservation. He believes the remaining 40,000ha could be best used by planting "a mosaic of small mixed species, mixed-aged forest blocks which would be beneficial for hen harrier protection".

This approach would also benefit merlin habitats, which are also under threat. John Lusby and fellow researchers in their study "Breeding ecology and habitat selection of *Merlin Falco columbarius* in forested landscapes" described a change in breeding habitat from open moorland to nesting in commercial forest plantations adjacent to moorland in Ireland:

> Although merlin nest in forest plantations, the presence of open suitable habitats in proximity to forest influences nest site selection and breeding success. Long-term trends in merlin breeding performance have not been negatively influenced by increased afforestation. However, in the absence of strategic monitoring, uncertainty over true population densities and trends remain.

The opportunities and challenges facing Irish forestry in supporting biodiversity are not unique in a global context. The decline of natural forests throughout the world has highlighted the role of plantation forests in biodiversity enhancement. Unlike most European countries, Ireland relies on plantation forests, which have potential to enrich biodiversity as outlined by Cormac J. O'Callaghan and fellow authors in the study "The role of planted forests in the provision of habitat: an Irish perspective". The reduced scale of forests and mix of species and open spaces provide opportunities to "take a broader look at the forest landscape as a whole to further improve the biodiversity value of plantation forests," according to the study, which also states:

Areas of Biodiversity Enhancement (ABEs) include trails, roads, roadside and river verges, forest rides and deliberately unplanted or understocked forest areas.

Areas of Biodiversity Enhancement (ABEs) are an important factor in the provision of habitat by plantation forests, and include a range of associated forest features such as roadside verges, forest rides and unplanted glades. Under current guidelines, approximately 15% of forest area must be treated with particular regard to biodiversity, including 5-10% open space and 5-10% retained habitat features.

This individualistic site by site approach to forestry acknowledges the changing structure and ownership of Irish forests. Therefore, a blanket banning of afforestation on hen harrier SPAs is counter-productive, as blanket afforestation is no longer an issue in Ireland's forestry programme. Farmers who work the land should have a say in the way land is managed, maintains Michael Fleming. His belief, that effective sterilisation of large tracts of excellent forestry land creates "a risk of land abandonment" among farmers "who have played a key role in protecting the hen harrier," is worth heeding.

FORESTS AND WATER

Globally, the relationship between water and forests is complex. While forests thrive in Ireland's moist temperate climate, they suffer stress in countries with drought conditions and forests may even reduce water availability by negatively affecting stream flows and catchment water yields. An understanding of the role of evapotranspiration in this dynamic is important. Evapotranspiration in the forest plays a key role in the water cycle, including water transpiration and interception, which affect water quality and water runoff (Figure 5.7). Evapotranspiration is the combined term for evaporation (water lost from the soil, ground vegetation and water surfaces) and transpiration (water lost from the surface of the plant). Trees have a larger leaf area than other vegetation as outlined by W.W. Verstraeten and co-authors and so "have higher transpiration rates, together with higher interception rates, resulting in trees having more evapotranspiration than other vegetation types".

The importance of rainfall interception by forest canopies varies in Ireland as precipitation fluctuates between 750mm in the east to 1,400mm in the west. Precipitation can exceed 2,000mm in mountainous areas particularly in counties Donegal, Mayo, Galway and Kerry. Irish forests can easily absorb this rainfall, which is fairly evenly spread over the seasons, in contrast with countries that experience drought conditions and widely varying rainfall patterns, although climate trends in Ireland are also changing. While rainfall in Ireland can support increased afforestation, the opposite may be the case in countries that experience prolonged drought. This is illustrated by S.A.O. Chamshama and F.O.C. Nwonwu in their report "Forest plantations in Sub-Saharan Africa":

Southern Africa is vulnerable to water stress and the situation is predicted to worsen, particularly in Botswana, Namibia and South Africa. In most of the dry areas of the region, high rates of

Figure 5.7: *Evapotranspiration – how trees intercept and transfer water in the forest cycle and reduce the risk of flooding.*

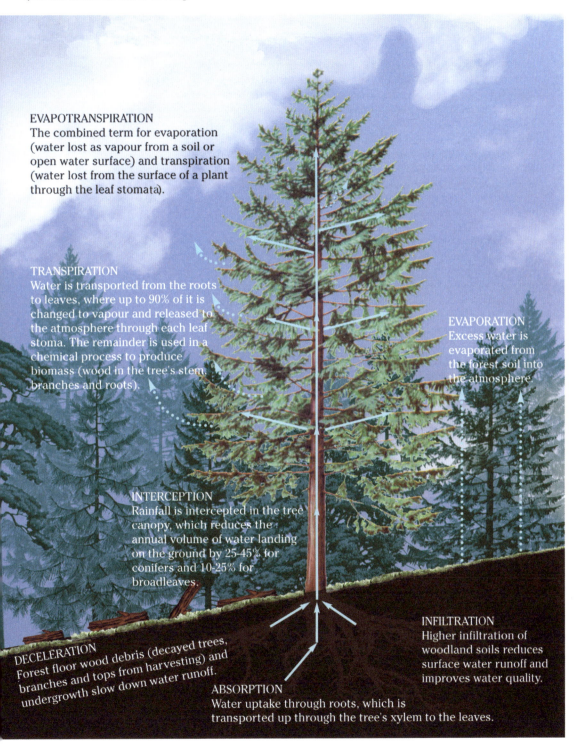

EVAPOTRANSPIRATION
The combined term for evaporation (water lost as vapour from a soil or open water surface) and transpiration (water lost from the surface of a plant through the leaf stomata).

TRANSPIRATION
Water is transported from the roots to leaves, where up to 90% of it is changed to vapour and released to the atmosphere through each leaf stoma. The remainder is used in a chemical process to produce biomass (wood in the tree's stem, branches and roots).

EVAPORATION
Excess water is evaporated from the forest soil into the atmosphere.

INTERCEPTION
Rainfall is intercepted in the tree canopy, which reduces the annual volume of water landing on the ground by 25-45% for conifers and 10-25% for broadleaves.

INFILTRATION
Higher infiltration of woodland soils reduces surface water runoff and improves water quality.

DECELERATION
Forest floor wood debris (decayed trees, branches and tops from harvesting) and undergrowth slow down water runoff.

ABSORPTION
Water uptake through roots, which is transported up through the tree's xylem to the leaves.

evapo-transpiration and competition for the limited water by other sectors of the societies (domestic, agriculture, industry) potentially limit plantation expansion.

In recent decades, prolonged drought conditions are also occurring in Europe along with severe flooding so water management is now an important aspect of forest management.

One third of European rivers flow through forested catchment areas, so the relationship between forests and water is interlinked in managing water with a two-fold aim: to maintain or improve water quality and to minimise flood risk.

Water quality

Scarcity of water worldwide has led to increased emphasis on managing forests to provide clean drinking water. More than 20% of European forests, mainly in mountainous areas, are dedicated to protecting water and soils.

Perceived negative associations between productive forestry and water, especially in relation to the planting of bogs and uplands, are being addressed in Ireland. Forests are no longer established on peats that are over 30cm deep while tree species composition, landscape planning and management of riparian areas are now designed to maximise the environmental benefits of forestry. Forest establishment and management adhere to a range of Forest Service mandatory guidelines in relation to water quality, landscape, harvesting and the environment, biodiversity and protection.

Ireland has more than 73,000km of river channels. The EPA states: "Three-quarters of these channels are very small streams that typically flow into larger rivers". Forestry has a major responsibility in ensuring water catchments provide healthy water while forests also have a key role in minimising flood damage.

The water purification role of forests, including that of forest soils, is stressed in the 2010 European Commission Green Paper. Many Member States implement their own water quality forestry regulations, including France, Germany, Belgium and Austria. In Belgium, water from the Ardennes Forest area is the principal source of drinking water for Brussels and Flanders.

In Germany, two thirds of the "Wasserschutzgebiete" – water protection area – is under forest cover to ensure high, quality drinking water. In Spain, forests in upper river catchments have been given special conservation status because of their capacity to improve water quality. Safeguarding the quality of drinking water is a key objective of foresters in Austria, a country which has one of the cleanest water resources in the world. The Austrian Forest Programme states:

> The city of Vienna with more than one and a half million inhabitants, for example, is supplied with freshwater from its own water collection forests.

Productive forests are particularly efficient at intercepting and reducing the delivery of diffuse pollutants to water from upslope land. For the nutrients, nitrate and phosphate, this mainly reflects the strong nutrient demand by the growing trees, which is maintained by regular thinning, while most forests are initially established or replanted when they mature without requiring artificial fertilisers.

Maintenance and enhancement of water quality features strongly in all aspects of sustainable forest management in Ireland. Since 2000 Irish foresters have complied with the Department's "Forestry and Water Quality Guidelines" which set out "sound and practical measures based on the principles of sustainable forest management and are firmly rooted in the best available scientific information".

These have been kept under review in the light of new developments as forestry has to address issues identified in Ireland's "River Basin Management Plan", "Environmental Requirements for Afforestation" and the "Draft Plan for Forests and Freshwater Pearl Mussel (FPM) in Ireland". The objective of the draft FPM plan, according to McCarthy Keville O'Sullivan Ltd. in 2018 was:

> to eliminate, reduce or mitigate diffuse pollution and maintain a more natural hydrological regime arising from forest activities undertaken within each of the 27 FPM catchments, to ensure that these activities do not threaten the achievement of the conservation objectives for the SACs involved, namely "to maintain or restore the favourable conservation condition of the Annex I habitat(s) and/or the Annex II species for which the SAC has been selected."

Forests and woodlands reduce the volume of rainfall landing on the ground by 25-45% on an annual basis for conifers (above) and 10-25% for broadleaves.

Opposition to new native woodlands in these catchments is environmentally damaging as these vital future ecosystems can be established and managed without fertiliser application, drainage, mechanisation or any soil or vegetation disturbance.

These catchments should feature permanent low management native and naturalised tree species that will protect FPM habitats.

Forest establishment needs to comply with guidelines laid down in "Woodland for Water" (Figure 5.8). This illustrates how "new native woodland and undisturbed water setback can be used in combination to deliver meaningful ecosystem services that protect and enhance water quality and aquatic ecosystems". The document outlines how well-sited, designed and managed woodlands and forests benefit water quality and aquatic ecosystems. They deliver a range of ecosystem services by preventing sediment and nutrient runoff, as well as erosion. This approach contributes to the shading and cooling of water, the overall restoration of riparian habitats and helps floodwater control. A practical example of how native woodland and undisturbed water setback actually fit into the landscape is illustrated in a 30m to 45m native woodland setback area which is created between the watercourse and adjoining land use which can include agriculture, commercial forestry or the built environment.

Figure 5.8. *Woodland for Water: planning to protect and enhance water quality and aquatic ecosystems.*

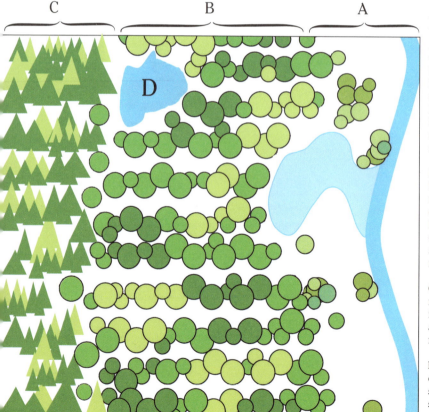

C   B   A

KEY

A: Permanent undisturbed water setback, 10-25m. Can be widened depending on site hydrology and slope. Uncrossed by new drains and largely unplanted except for individual or small groups of native riparian trees and shrubs.

B: New native woodland at least 20m wide uncrossed by new drains and planted according to the Native Woodland Scheme. Width dependant on adjoining land use, site hydrology, slope and vulnerability to receiving waters.

C: Adjoining land use: commercial conifer forest in this illustration. Could also be agriculture or built environment.

D: Potential to block existing drains with silt traps and slow-flow dams and create small pockets of wetlands and settlement areas surrounded by species such as alder and willow.

Source: Adapted by D. Magner *from Woodland for Water: Creating New Native Woodlands to Protect and Enhance Ireland's Waters*. Department of Agriculture, Food and the Marine, 2018.

RAVEN – LANDES: THE FRENCH CONNECTION

Given the vulnerability of the Raven, Co. Wexford to sea erosion, the main aim was to protect the dunes from washing away when planting began in 1929. The history of reclamation in Wexford Harbour began in 1846 when John Edward built a sea wall to reclaim almost 1,000ha of harbour from the sea but 80 years later a series of storms washed away the southern sand spit.

To protect the sand dunes, it was decided to plant a mainly pine plantation (top) beginning in 1929, which was influenced by the Forest of Landes stretching inland from the Bay of Biscay in southwest France (bottom). Maritime pine is the species common to both man-made plantations but the Irish foresters of the time also added Corsican pine, Scots pine, Douglas fir and Sitka spruce. They even experimented with monkey puzzle or Chilean pine and sycamore because they are tolerant of salt-laden winds.

The role of forests in preventing soil erosion and flooding

Globally, sustainably managed forests play an important role in preventing soil erosion and flooding. In Europe, the combined benefits of forestry in mitigating flood damage and preventing soil erosion have been acknowledged for centuries. For example, in France, the Restoration of Mountain Lands – *Restauration des Terrains en Montagne* – was established to protect land after the floods of 1856 and 1859 which resulted in part from deforestation. Erosion prevention, soil improvement and flooding abatement were key objectives when the largely man-made *La forêt des Landes* – the Forest of Landes – was created in southwest France in the eighteenth century using mainly maritime pine. This resulted in the fixing of the sand dunes from the encroaching sea around the Bay of Biscay. Further inland this grand afforestation experiment resulted in flood prevention and soil amelioration in the vast infertile moors. Eventually some of the forested area was converted to agriculture, which allowed rye and maize cultivation in once impoverished soils.

Augustine Henry delivered a lecture in the (RDS) Royal Dublin Society in 1904 after visiting the Landes where he said, "a desert had been made healthy and productive". Henry's experience of the Landes influenced pine planting in a number of coastal forests in Ireland including the Raven in Co. Wexford.

Most European countries are tackling flood control through a combination of technology such as flood barriers and sustainable management of both upland and riparian woodlands. This approach is emphasised in the European Commission's 2010 Green Paper "Preparing forests for climate change":

> Forests play a major role in the storage, purification and release of water to surface water bodies and subsurface aquifers. Their soils buffer large quantities of water, reducing flooding.

Productive and non-productive forests play a key role in reducing water flow as outlined by Tom Nisbet in *The Role of Productive Woodlands in Water Management*. The benefits identified by Nisbet in reducing water flow include:

- Greater "water use" of trees reduces the volume of water at source.
- High infiltration rates of woodland soils reduce rapid surface runoff and flood generation.
- Greater hydraulic roughness exerted by trees, shrubs and large woody debris slows down flood flows and enhances flood storage.
- Ability of trees to protect the soil from erosion and interrupt the delivery of sediment runoff helps to maintain the capacity of river channels to convey flood waters downstream and reduces the need for dredging.

Nisbet maintained that coniferous forests provide larger evaporation or 'interception' of rain water than broadleaves.

His findings show that conifers can reduce the volume of rainfall landing on the ground by 25-45% on an annual basis, compared with 10-25% for broadleaves. Both conifers and broadleaves have a major role in regulating water flow and improving water quality as well as other environmental benefits as outlined by David Ellison and co-authors in their 2020 paper:

> Tree and forest root systems contribute to better soil water infiltration and their large leaf litter production leads to more soil carbon, which improves both water retention in the soil and groundwater recharge.

While Nisbet and Ellison agree that coniferous forests provide greater benefits in reducing flooding, Ellison explains that a different approach in tree species selection is necessary for water quality improvement, especially in areas where air pollution is an issue. Broadleaf forests are preferred in drinking water production areas, he claims, because their "lower average leaf area yields more water in the aquifer, and because this water is also less contaminated, as conifers intercept more atmospheric pollution". While the obvious solution is to reduce air pollution, forest management practices can incorporate pure broadleaves, especially as buffer zones in riparian areas as outlined in Figure 5.8 (p104). In these catchments, forestry has inherent advantages over other land uses in enhancing water quality. For example, chemical run-off can be avoided as there should be no requirement to use fertilisers, herbicides, fungicides or insecticides.

To summarise, the role of forests and water, research results demonstrate the benefits of forests in both flood control and improvement in water quality when the correct forest management procedures are practised. This is acknowledged in Ireland's Climate Action Plan (2022):

> Increased forest cover can retain excess water and mitigate the impacts of floods, increasing resilience to climate change.

FORESTS AND PUBLIC GOODS

While wood production is generally the primary objective in forest management, forests are also the source of many important non-wood products and services, also known as public goods. In some countries these are as important as wood production. Globally, "forest ecosystem services support agricultural, livestock and fishery production through water regulation, soil protection, nutrient circulation, pest control and pollination," according to the FAO. Our forests' non-wood services such as tourism and recreation have provided major social and economic benefits since the State adopted an open forest policy in the 1970s.

The delivery and funding of public goods is an essential component of future forest planning and policy as forest owners

TREES, FORESTS AND INTEGRAT

The natural forests are long gone in the Anne Valley in east Co. Waterford but trees and woodlands play a major role in this intensely farmed area. Water is the unifying component, as the Annestown stream is the corridor that connects wetland, farm land, wildlife habitats, forests and woodlands.

In 1992, Dr. Rory Harrington, a scientist and forester, convinced 12 farmers in this catchment that nature rather than costly technology would be the solution to water pollution. He proposed transforming water treatment by establishing the Integrated Constructed Wetland (ICW) concept.

ICWs comprise a series of interconnected wastewater treatment systems or ponds planted with sedges and reeds. These purify wastewater and sewage from farms and dwellings before returning it to Annestown stream.

The project was supported by the Waterford LEADER Partnership, National Parks and Wildlife Service, and the Office of Public Works. ICWs are licensed under the EPA discharge licence system and Irish Water is responsible for their maintenance.

Dr. Rory Harrington (left), initiator of Dunhill ICWs, forestry consultant Dermot Dunphy and Dunhill farmers – Niall Moore and Willie Moore, who along with 10 other farmers implemented the scheme.

NSTRUCTED WETLANDS (ICW)

The wetland ponds and their vegetation are central to the ICWs but Harrington and the farmers have ingeniously combined these with a series of forest corridors and small woodlands. These not only act as natural buffers but also enrich the flora and fauna, enhance the landscape and are now used by the public as a recreation and educational resource.

He recognises the importance of trees – native and introduced – in the project. Native species such as ash, willow, whitethorn and alder populate the bank of the Annestown stream. Dr. Harrington also enthuses on the wide range of species that grow well in Ireland including exotics such as Sitka spruce, Douglas fir Monterey pine, coastal redwood and southern beech.

He maintains that projects like the Dunhill ICW system don't require large-scale transfer of agricultural land in providing clean water and climate change mitigation benefits. The combined annual wetland CO_2 sequestration potential from Ireland's farmyards alone could be up to 4 million tonnes based on no more than 2ha of functional wetland area per farm.

Annestown stream catchment showing the re-profiled stream, wooded corridor accessible for tractor-based maintenance and linked wetlands that intercept and treat water runoff from adjacent fields.

currently receive no remuneration for providing these societal and environmental benefits.

Public goods can be divided into non-wood environmental services and non-wood products. We have already explored the role of forests in the provision of non-wood environmental services including forests and water, landscape, recreation, biodiversity and carbon trading. In recent years, the role of forests in carbon sequestration is being addressed, especially in Europe. This is demonstrated in the new EU Forest Strategy which seeks *inter alia* to develop, "financial incentives, in particular, for private forest owners and managers, for the provision of these ecosystem services". This approach acknowledges the ecological and economic benefits of forestry, which is important in Ireland as forest owners are expected to establish at least 50% native species as outlined in the government's forest strategy.

In addition to non-wood services, forests provide a range of non-wood goods which are mainstream in some European and global forests but are only in their infancy in Ireland.

Forest foraging

Forests provide edible plants, fruits, nuts, seeds, fungi and edible oils for humans as well as fodder for both meat and dairy animals. In Ireland we have only begun to explore the forest as a source of food, especially fruit and edible mushrooms or fungi.

Comparisons with Scandinavian countries, where there is widespread use of non-wood produce, are worthwhile. This is not deterred because of coniferous forest dominance in countries such as Finland, where the tree species mix is 89% conifers – mainly Norway spruce and Scots pine – and 11% broadleaves, almost all birch. "Approximately 500 million kg of berries and a staggering two billion kg of mushrooms grow in Finland's forests every year and the tradition of picking wild berries and mushrooms is as popular as ever, despite urbanization," writes Salla Korpela. In the heart of the boreal coniferous forests in the region of Kainuu she claims "each household picks an average of almost 60kg of berries a year".

The European tradition of foraging for wild bilberries, raspberries, loganberries, cranberries and mushrooms could be matched in Ireland, as most of these species grow well here. Despite the relative youth of the Irish forest estate, mushrooms are produced in abundance. Paul Dowding and Louis Smith feature 70 species in *Forest Fungi in Ireland*. These comprise 43 edible, 14 inedible and 13 poisonous species. Most of these are found in coniferous forests and as a result, mushroom foraging is becoming increasingly popular.

Foraging is likely to be a major non-timber forest use in future partly due to the presence of edible fungi in mixed species and native woodlands, but especially in our predominantly coniferous forests. I noticed how popular foraging is among European migrants during my visits to Ireland's forests while researching *Stopping by Woods*. Foraging is especially popular among eastern Europeans who have set up home in Ireland and who value the forest as both a food and wood provider. Eoghan Daltun takes a different view of foraging:

> I just prefer to leave it all alone. We are not living in hunter-gatherer times, when tiny bands of people roamed vast, barely touched landscapes; totally the opposite.

Daltun is referring to his own valiant rewilding project on the Beara Peninsula, which he recorded in *An Irish Atlantic Rainforest*, but if we are really to succeed in getting people to connect with the forest ecosystem, surely foraging is one of the most innovative and least damaging ways of achieving culinary, ecological and educational objectives.

Mary Bulfin, wild food forager, chef and author demonstrates a range of food products she sources in forests and woodlands in the midlands. She is seen here in the sustainably managed Clonad Forest with certification forester Sarah Standish, Irish Forest Unit Trust.

While most foraging focuses on the plant and shrub layer of the forest, tree nuts, seeds, fruits and leaves are a rich source of food, preserves, additives and drinks. For example, fresh Norway and Sitka spruce needle tips are used to flavour craft beers, while beech leaves are an ideal base when making brandy or gin-based liqueurs. Edible nuts are produced by the native hazel and naturalised sweet chestnut.

Trees have been explored for millennia, not just as food sources but also for their medicinal properties as illustrated in the ninth-century poem, "Marbhan and Guaire" (translated by Paul Muldoon):

> *Hawthorn good for a pain in the heart.*
> *Yew for giving it a start.*
> *Blackthorn tea for a medicinal rinse.*

Farm forestry

When the EU adopted a policy to decrease food production in the 1980s, Irish farmers were encouraged to switch land use to forestry. During the period 1980 to 2022, 23,859 landowners – mostly farmers – planted land. Farmer planting exceeded 85% of all planting over the period. While encouraging, this

Native Tree Area scheme is a flexible afforestation initiative suitable for small woodland blocks including difficult field corners, wet areas and fields close to rivers and streams such as this site in Moorepark, Co.Cork beside the River Funshion.

represents only 16% of 135,000 farm holdings. There are many reasons why farmers are reluctant to plant, not least the legal replanting obligation, long-term investment and environmental constraints. Forestry is also seen as a large land use that could engulf Irish farms, which average 33.4ha even though forests established in Ireland during the period 2013 to 2023 averaged 7.5ha. One way of getting farmers to explore the forestry option is to begin by establishing small-scale forests and woodland. Two schemes – Native Tree Area (NTA) and agroforestry are tailor-made for farmers who are unsure about converting large areas of land to forestry at the expense of conventional farming (see also Chapters 1 and 9).

Even good agricultural land has small areas that could be planted, but up until recently afforestation schemes were prohibitive for small lots. The NTA scheme is suitable for all farms as it allows planting areas as small as 0.1ha or a quarter of an acre without requiring time-consuming and costly licences. In all, 2ha can be planted under the NTA scheme.

While livestock are generally regarded as incompatible with forest management objectives, mature forests can provide shelter and grazing for cattle and sheep, while agroforestry is an option for farmers to combine forestry and agriculture. Agroforests are an ideal habitat for sheep, pigs and cattle depending on crop age. The wide spacing of trees also allows cropping for silage, hay and cereals.

In Ireland and developed countries, agroforestry is fully compatible with agricultural activities – crops, pasture and livestock – and timber production. In tropical countries it has wider aims including erosion prevention, shelter and soil replenishment.

Agroforestry is still rare in Ireland but is not unusual in continental Europe. For example, in southeast France, wheat, lavender and other crops are grown between rows of trees such as walnut and poplar. Nor is agroforestry confined to forest trees: fruit crops such as apple trees combine well with traditional farming crops in the US, while in Canada intercropping with strawberries is carried out in peach orchards.

On suitable holdings, agroforestry is recognised as an alternative to full forest cover. It allows mainstream farming to take place alongside forestry and has the potential to produce high-value timber crops. It is particularly suitable where trees and pasture are combined, so it is an option in Ireland as livestock are present in over 90% of Irish farms.

Agroforestry has potential for a limited range of conifer species, but broadleaves are regarded as more suitable. While ash is ruled out because of disease vulnerability, suitable native species include cherry, alder and oak. It is also suitable for non-native broadleaves especially sweet chestnut, sycamore, poplar and walnut.

One of the major advantages of forestry is its compatibility with agriculture. Even with 60% forest expansion by 2050 to achieve carbon neutrality, forestry will not negatively impact on agricultural production so Irish farmers have both the land resource and climate to allow compatible development of forestry and agriculture. Non-wood goods and services can best be achieved in conventional forestry and forest rotations but forest owners also need to be aware of, and open to, other silvicultural practices including continuous cover forestry and agroforestry.

While native woodland and agroforestry schemes are attractive to farmers who wish to retain most or all of their holdings in agriculture, many opt for commercial high forests which provide a strong revenue stream after State premiums cease. Tree species selection is key in these areas, with the greatest return from commercial coniferous crops (see Chapter 8).

FORESTRY AND THE RURAL ECONOMY

Forestry plays an increasingly important role in the rural economy mainly through the provision of sustainable employment in forests and downstream industries, and diversification of farm incomes and tourism. The annual economic value of forestry is estimated at €2.3 billion, virtually all of which is generated – and stays – in rural economies (Table 5.9). Most of this is generated from productive commercial forests but forest recreation also brings tourism benefits to rural communities. For example, the total annual economic activity generated by domestic forest users in Ireland is estimated at €268 million while walking tourism accounts for €138 million.

Employment in the forest and wood products industry is rurally based, where jobs are badly needed in an increasingly urbanised society. These jobs are spread over a wide range of activities including:

Forestry is a major employer in rural areas, beginning with plant production in the nursery. None so Hardy Nurseries in Counties Wicklow and Wexford produce 20 million plants annually over three-year cycles, beginning with seed sourcing including acorn collection in Coollattin Wood by Patrick Kilbride and Patrick Hennessy (opposite).

Below: some of the 80 nursery staff employed by None so Hardy in Donishall, Co. Wicklow and Ballymurn, Co. Wexford.

Table 5.9: *The annual value of forestry, comprising growing, harvesting, wood processing and related activities.*

	2003	2012
Growing and harvesting subsector	million €	
Direct Economic Activity Value	255.4	386.9
Gross Value Added	134.5	136.6
Overall Value to Wider Irish Economy	472.4	688.7
Type 2 Multiplier	1.9	1.78
Wood processing subsector		
Direct Economic Activity Value	975.0	1,389.1
Gross Value Added	312.3	391.6
Overall Value to Wider Irish Economy	1,650.0	2,299.4
Type 2 Multiplier		1.66

Source: Forest Statistics Ireland 2021, Department of Agriculture, Food and the Marine.

- Forest nursery – seed collection, plant production and tree improvement research programmes.
- Forest management – ground preparation, cultivation, fencing, planting, road construction, fire protection and forest maintenance.
- Woodland improvement – tending, thinning and pruning.
- Production – commercial harvesting throughout the crop rotation.
- Reforestation – establishing a new crop after final harvest or conversion to continuous-cover forestry.

These activities begin at seed collection and the production of nursery plants. Approximately 60 million nursery seedlings will be required annually to plant 25,000ha comprising at least 10,000ha afforestation and a further 15,000ha reforestation, if Ireland is to achieve a viable forestry programme. The main activities are at the establishment phase, where operations include ground preparation, fencing and planting followed by early maintenance to ensure the crop is free of vegetation competition and is protected against fire, disease and mammal damage. Fast-growing conifers and some broadleaves establish quickly while slower-growing broadleaves such as oak and beech require greater maintenance and care to ensure their establishment. State grants and premium payments are designed to reflect this silvicultural challenge.

The employment generation of a forest is high during the establishment phase and again during the tending and production phases. Based on COFORD estimates for employment creation from forestry, a 10,000ha annual planting programme would require 327 direct jobs at establishment, maintenance, management, harvesting and wood processing and downstream industries including:

- Processing sawn timber for construction, fencing and pallet (packaging) markets.
- Processing wood residue (wood chips, sawdust, etc) into panel boards including medium density fibre boards (MDF), oriented strandboard (OSB) and moulded door facings.
- Energy – conversion of waste timber residue to wood chips and wood pellets for the wood energy market.
- Other products – including furniture (interior and exterior), craft, panelling and turnery.

Currently 10,000 jobs are generated in forestry while 23,000 landowners, predominantly farmers, are involved in the maintenance and management of their own forests. If a 10,000ha annual afforestation programme is achieved up to 2050, forestry would create a further 9,800 jobs based on COFORD estimates as well as providing a sustainable income for an additional 30,000 landowners – mainly farmers, if they choose to plant an average of 10ha on some of their holdings. Therefore, by mid-century, forestry has the potential to provide up to 20,000 jobs in Ireland and an income stream for up to 53,000 forest owners, predominantly farmers.

Alder transplants at None So Hardy Nursery, Ballymurn, Co. Wexford.

These projections depend on future State commitment to afforestation and demand for timber and timber products in the domestic and export markets. The State has clearly outlined the crucial role of forestry in Ireland's Climate Action Plan and its role in achieving carbon neutrality by 2050. There are sufficient benefits to justify forests – and increased afforestation – by emphasising their non-wood benefits alone, which I have addressed in this chapter, But the Climate Action Plan has gone beyond the forest to fully justify its role in future sustainable living. It recommends industrial wood usage and engineered wood such as "cross-laminated timber and timber frame that can replace concrete and steel in many applications such as floors, roofs, walls and stairs due to their strength and versatility". Future demand for timber and timber products is forecast to increase in Ireland, Europe and globally.The market is driven by increased construction demand coupled with climate change policies to decarbonise economies, as Richard van Romunde, points out in *Global Timber Outlook 2020*:

> Globally, the vast majority of countries have set significant targets to reduce carbon emissions towards net zero by 2050. Timber will play a critical part in this transformation. The dual effect of urbanisation and decarbonisation will be more new homes and cleaner low carbon intensity buildings being built from timber.

While I emphasise the non-wood benefits of forests in this chapter, there is no escaping the role of commercial forest owners as the providers of timber that will displace fossil based materials in construction and energy, both massive global and domestic greenhouse gas emitters. Their role is acknowledged in Ireland's Forestry Strategy 2023-30:

> For the majority of forest owners selling timber is the primary
> source of income from forests and these products will be key
> in Ireland's transition to a sustainable climate neutral economy.

Not to address the topic of wood and wood products would be to deny the totality of forestry, its diversity and the its centrality in living, which I explore in the next chapter. To address the multifunctionality of forests without addressing its multipurpose end uses would be to deny sustainable construction, packaging, furniture, energy and print. It would be a rejection of current and future research into new wood-based products including fabrics, food, liquid fuel and even microchips as well as the exciting possibilities of engineered wood and reuse. Above all, it would be a denial of wood.

CHAPTER SIX

Why Wood?

Everything that's made with fossil fuels today can be made from a tree tomorrow.
Karl-Henrik Sundström, 2017

Timber cities would lock up a great deal of carbon. Even more to the point: although it requires energy to turn a tree trunk into a finished beam...it takes roughly twelve times as much to make a steel girder that is functionally equivalent.
Colin Tudge, *The Secret Life of Trees: How They Live and Why They Matter,* 2005

Using wood and wood-based products for construction is a sustainable substitute for conventional carbon-heavy construction products, such as concrete, brick and steel.
Climate Action Plan: Securing Our Future, 2021

Perhaps it is the call of the past that brings wood back into architecture; maybe it is ecological and economic sense, or perhaps wood is a material that is closer to our own nature than steel or concrete.
Jodidio, Philip, *100 Contemporary Wood Buildings,* 2017

Wood is enjoying a renaissance as a major contemporary construction and design material. Innovative architects, engineers and designers are once again turning to wood as the medium of choice. They are reconnecting with wood because of its aesthetic, durable, versatile and tensile characteristics. In addition, timber's renewable and ecological qualities have placed it at the heart of twenty first century sustainable living; where the traditional and the contemporary meet, and where questions are being asked of all materials and their impact on climate change and our environment.

Wood never lost its aesthetic, creative and functional expression in furniture, joinery, panelling, craft, music, sculpture and a wide range of domestic applications. It also maintained its worth in construction but declined as a material in large-scale projects, which began during the Industrial Revolution and accelerated in the nineteenth century. The most dramatic example of its obsolescence occurred in shipbuilding during the 1860s as the new ironclad battleships destroyed the wooden fleets in European naval battles and in the American Civil War. Almost overnight, a key forest management objective – to produce oak timber for the world's powerful navies – was abandoned. While not as abrupt, the decline of wood in construction continued apace. As structures became larger, wood was replaced with steel and concrete. Paradoxically, timber's role in displacing these non-renewable is now a key to its emergence as a contemporary renewable construction material.

THE STORY OF WOOD

An understanding of the rise and fall and re-emergence of wood is important in realising its future potential. Fortunately, we have both archaeological and physical evidence of wood usage over thousands of years to enlighten us on its past significance and the new directions wood could take us.

The world's great engineering, architectural and design achievements began with wood, in buildings, ships and bridges, while our domestic and social history is shaped by wood. It was the first material to provide shelter, furniture, weapons and agricultural implements. It found cultural expression in the making of music, craft and art.

Yet, evidence of timber artefacts is scarce compared with other materials, which may be the reason why there is no Wood Age. Although it is perishable, there is ample evidence that wood was used as tools and weapons, long before the Stone Age. Many archaeologists have dismissed wood – and a Wood Age – too readily, opting for "the traditional, outmoded but convenient Three Age system of European Prehistory", claim J.M. Coles and fellow authors. They argue that "there is hardly a tool or weapon used by Stone-Age, Bronze-Age or Iron-Age man or woman which did not have a wooden part ...". In predating the Stone Age and straddling all Three Ages, wood demonstrated not just its versatility but also its over-familiarity; too easy to discard and too ordinary to warrant serious study. Wood deserves deeper study than it has been afforded to date, to get a greater understanding of the part it played down through the millennia from its early use in prehistoric communities to its re-emergence as a vital medium in twenty-first century sustainable living.

The generic terms "wood", "timber" and "lumber" (used mainly in North America) cloak the myriad of characteristics of individual tree species. The modern timber supplier probably has no more than a dozen different species at hand, most with no linkage to ancient Irish species. Yet, all are still growing here and once were embedded in the psyche of whole communities. Take yew (*iúr*, in Irish) for example, with its multifaceted qualities including longevity, spirituality and durability. These characteristics provide it with an aura above and beyond other more prominent tree species today. Why else did communities adopt it as their *genius loci* in places such as Terenure, (*Tír an Iúir* – place of the yews), Virginia (*Achadh an Iúir* – field of the yew) or whole landscapes in Mayo (*Maig Eo* – plain of the. yews)? The tree of church and graveyard provides clues of its spirituality that transcended the ancient Celtic and Christian but in a practical sense woodworkers have long valued yew as the wood of choice for weaponry including longbows and arrows. So, when a 420,000-year-old pointed yew shaft was discovered in Clacton-on-Sea, Essex in 1911, archaeologists wondered about its original function. It may have been a digging implement, stave, snow probe or lance, but the evidence and the historic weight of yew's warrior history trumped the commonplace. So, the archaeologists catalogued it as the Clacton Spear, leaving little doubt that its distant role lay in battle or hunting.

Clacton Spear, 367mm x 37mm, Natural History Museum, London, c420,000 years old.

Early hunter gatherers had little interest in the large-scale wood use from the vast forests. Instead, they fed off fish, game, fruits and nuts and the little wood they used was for weaponry and crude shelters. The history of the first buildings in wood is uncertain. In Europe, there is evidence of wood use in construction in the Mesolithic period. During 4410-4325 cal BC, calibrated radio carbon dating, suggests that wood may have been used in Orkney for "construction purposes or as a marker/possible landing place showing the path to the Loch of Stenness" according to Scott Timpany and fellow researchers.

The first major timber clearances in Ireland took place around 3750BC with the arrivals of the first farmers who felled large trees for agriculture and more sophisticated buildings and boats. The hollowed-out oak canoe in the National Museum illustrates that boat-building was alive and well in Ireland from at least 2500BC (see Chapter 4). Now, rivers were no longer barriers to inland exploration but bridges were also required to traverse what was still a thickly wooded landscape.

Wood was used in Europe for bridge construction at least 3,500 years ago. The piles in the Holzbrücke Rapperswil-Hurden bridge on Lake Zürich, Switzerland date to 1523BC. There is no evidence of bridge-building during this period in Ireland but it's likely that the first experiments were taking place to cross rivers at strategic points during this period.

The Corlea Trackway, near Kenagh, Co. Longford, is a fine example of wood use in constructing crossings over other great watery expanses such as bogs. The deep raised bog in this part of Co. Longford would have been too difficult to traverse on foot or horseback, so a 1-km long land bridge was constructed to connect an upland area in the east to a small dry island in the west. According to dendrochronological analysis, the oak was felled in 148 or 147BC, then planked and laid down on the bog. The road's purpose – functional or ceremonial – is unclear. In any case, the builders hadn't adequately researched oak's suitability as a bog road. The species is durable but has a high density and would have been especially heavy when it was laid as it wouldn't have air-dried sufficiently. Within a short period, it sank without trace into the bog where it was preserved for over two millennia before it was discovered in 1984 during peat milling by Bord na Móna. An 18-m stretch of the road is on permanent display in the purpose-built Corlea Trackway Visitor Centre. Like the earlier boat builders, the Iron-Age people around Corlea demonstrated their ability to cut, shape and transport large oak trees even thought the end result was a failure. The Corlea Trackway involved the felling of 300 large oak trees, which also demonstrates how quickly a forest could be denuded either for construction or agricultural clearance.

The first evidence of large-scale bridge building in Ireland dates to c.AD804, when a 120-m wooden bridge was constructed on the River Shannon, downstream from the Clonmacnoise monastic site. Documented in a 1997 study by Donal Boland and Aidan O'Sullivan, the bridge includes a double row of eroded wooden posts at water depths of 1.5 to 5m. Still visible underwater today, the bridge is up to 5m wide in places. There are other examples of wood use in Ireland such as fencing around livestock enclosures. In Co. Cork, fencing walls in a Middle Bronze Age enclosure (1700-1500BC) "were formed by split oak timbers (c. 0.1m thick)", as recorded by Ken Hanley following excavations during road construction in 2005.

While we can trace the evolution of wood use through the millennia, early evidence can be uncertain, as wood is subject to fungal decay and unless the right soil and atmospheric conditions exist, the evidence is lost, especially in the case of large-scale construction projects. However, a number of wooden buildings over 1,000 years old still survive, as outlined by Marian Moffett and coauthors of *A World History of Architecture*.

An 18-m stretch of the bog road (above and opposite), dating to 148BC, is on permanent display in Corlea Trackway Visitor Centre, near Kenagh, Co. Longford.

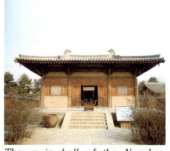

The main hall of the Nanchan Temple, China, 782AD, is still in use.

The Nanchan Temple in China – built in 782AD – is still in use, while a number of buildings in Europe have survived since the twelfth century.

In Norway, the stave church at Urnes, dates to c1130, while the oldest secular building is a storehouse on the grounds of West Telemark Museum, dating to 1167. The Granhults Church, 30km northeast of Växjö, is the oldest entirely preserved wooden church in Sweden, with the oldest sections dating to the early 1200s.

Once the ancient builders could shape and transport stone, wood began to lose its central role as a major building material. Stone's monumentality and long-lasting characteristics are still in evidence in iconic structures such as Inca architecture, the Pyramids in Egypt, Stonehenge in England and Newgrange, Co. Meath, which predates these edifices. The Newgrange passage tomb which dates to c3200BC illustrates how Neolithic builders found a way not only to build in stone but also to transport it by river and sea. Forests would have been extensive in Ireland during this period, so wood played a part in the transport of quartz and granite, by boat or float. Wooden poles were used in stone logrolling at source and from the bank of the River Boyne, and finally in timber scaffolding during construction.

With the evolution of a European and Middle East masonry culture, wood was combined with stone. The Basilica of Constantine at Trier, constructed c310AD, is an example of this hybridisation where, according to Pao-Chi Chang and Alfred Swenson, "king-post roof trusses (triangular frames with a vertical central strut) span a hall 23m (75 feet) wide" and while the present roof is a restoration, they believe the original "must have been similar".

Wood was still required in major buildings but it was losing ground to stone especially in the late twelfth century which heralded the demise of wood as a major construction element,

The traditional storehouse originally located in Vindlaus, Norway, dates to 1167. It is now in the Museum of West-Telemark in Eidsborg, Norway.

especially in church architecture. To build high in stone, church architects faced major challenges in overcoming problems of stability and weight. These problems were solved during the construction of the Cathedral of Chartres (1194-1220) as described by Kenneth Clark:

> Now by the devices of the Gothic style – the shaft [of the cathedral] with its cluster of columns, passing without interruption into the vault and the pointed arch – [man] could make stone seem weightless: the weightless expression of his spirit.

At 105m tall, Chartres proved that architects could build high in stone, but wood – especially oak – was still used in churches such as Notre-Dame Cathedral, which relied on oak supports for its original 78m spire. After the spire was damaged by wind and decay, it was taken down in 1792 and replaced by a 96m spire, which was unveiled in 1859. Ironically, the oak forests planted during the nineteenth century, but no longer required by the French Navy, are now being sawn and dried to rebuild the Notre-Dame spire, destroyed by fire in 2019.

Although wood lost further ground as a major construction medium, it was still required as a versatile material in many structures during the Renaissance, as outlined by Pao-Chi Chang and Alfred Swenson:

> In addition to Roman forms in masonry, the Renaissance recovered other Roman technologies, including timber trusses. Giorgio Vasari used king-post timber trusses for a 20m span in the roof of the Uffizi or municipal office building, in Florence in the mid-16th century. At the same time, the Venetian architect Andrea Palladio used a fully triangulated timber truss for a bridge with a span of 30.5m.

While wood was relegated by stone in major European architectural projects, it continued as the main building

The Granhults Church, 30km north-east of Växjö, is the oldest entirely preserved wooden church in Sweden, dating to the early 1200s.

material in countries with strong wood cultures. These include Norway, Sweden and Finland in Europe and China and Japan in the Far East where large and medium construction in wood still flourished during the fifteenth century. For example, the Forbidden City, Beijing, built between 1406 and 1420 and the five-storied Yasaka Pagoda, Kyoto whose lineage dates to the sixth century was rebuilt in 1440. Like Scandinavia, China had an abundant forest resource. Chinese fir (*Cunninghamia lanceolata*) and nanmu (*Phoebe zhennan nanmu*) were the main species used in building the Forbidden City. Some of this vast complex has been damaged by fire over the centuries and rebuilt but remarkably it has survived an estimated 200 earthquakes. Sadly, many of the forests used in wood construction were unsustainably harvested and did not survive.

Wood continued as a secondary building medium throughout Europe from the fourteenth century but still featured as an essential material in hybrid stone-timber construction. Despite degrade, it is a relatively easy material to restore. The present hammer-beam oak roof in Westminster Hall, constructed in 1393, replaced the original eleventh-century structure.

Today's architects and engineers are reluctant to fell scarce oak to either build or restore but sometimes this is unavoidable in order to maintain the continuum of the original structure as is happening with the major Notre-Dame Cathedral restoration project. On a more modest scale, when the roof of Carrickfergus Castle, Co. Antrim required restoration in 2017, the architectural, construction and conservation team had no option but to choose oak to maintain the integrity of the original twelfth century castle. Luckily, they were able to source the principal trusses of the new roof from Irish oak felled by Storm Ophelia in 2017. Completed in 2020, the restoration team avoided the use of nails, screws and other metal material in securing the trusses with "green" oak-pegged fixings (see also "Restoration and Conservation").

Wood maintained its structural and aesthetic presence up until the Industrial Revolution, which signalled its demise as architects and engineers placed greater emphasis on building larger and higher as well as constructing bridges to span wide rivers and gorges. The Menai Bridge (*Pont Grog y Borth*) in Wales, originally constructed in stone, wrought iron and cast iron (1819-1826) dispensed with timber except for the wooden surface which was replaced with steel in 1893. Thomas Telford's design is not only a metaphor for a second Iron Age but also political division as it was built mainly to accommodate the increased traffic between Ireland and Britain as result of the Act of Union in 1800.

Materials such as cast iron, and later steel, provided the opportunity to build tall. The arrival of prestressed concrete and steel revolutionised construction and wood was left in their wake as a structural material for large scale building projects.

Nowhere was this construction revolution more in

WOOD IN EARTHQUAKE PRONE AREAS

Like the Forbidden City and the Yasaka Pagoda, other ancient Chinese and Japanese buildings also survived numerous earthquakes, which is why architects are re-examining wood as a construction material in earthquake prone regions.

Engineered wood construction has been explored by the University of Canterbury (UC), New Zealand, which began a collaboration project with Pre-Stressed Timber Limited on the use of Pres-Lam a pre-stressed laminated timber following the disastrous earthquakes in Christchurch in 2010 and 2011.

Pres-Lam can be used in conjunction with other engineered wood products such as glulam (GLT), laminated veneer lumber (LVL) and cross-laminated timber (CLT). UC Civil and Natural Engineering faculty developed a timber construction system which is designed to minimise damage during an earthquake. Since research began, 10 buildings have been built in New Zealand and Japan utilising this technology including UC's Beatrice Tinsley building (pictured), completed in 2019.

The Carrickfergus Castle roof restoration project used Irish oak felled by storm Opehlia in 2017.

evidence than in the US. Within a few years in the 1880s, wood became a memory as steel showed the way, horizontally in Brooklyn Bridge (1883) and vertically when the first skyscraper – the Home Insurance Building – was built in Chicago (1889), the same year as the Eiffel Tower.

Wood continued as an important element in domestic small-scale buildings, especially in Nordic countries and North America. From the early nineteenth century, the US enjoyed a revolution in small-scale timber-building due mainly to the abundance of coniferous forests and the advent of mechanised sawmills as described by Robert Hughes:

> Take a tree and run it through a mill, and you have the first element in American prefabrication, the dimensionally accurate stick of lumber. Standard lumber sizes – two by four, two by six and so on – enabled builders to plan and cost out structures with an accuracy that had not been available before, in the days when wooden frames were made of thick baulks of timber, hand-sawn, adzed, mortised and tenoned on the site, and joined with wooden pegs. Dimensioned mill sawn lumber, fixed with the wire nails that the Eastern factories were turning out, enable anyone of ordinary practical skills with hammer, saw and rule to knock together a house anywhere in America.

These "balloon frame" houses described by Hughes, depended on an abundance of long, straight logs. They could be erected quickly but had design faults which resulted in fires and degradation. They were gradually replaced by timber frame houses, which require shorter and smaller logs. These dominate domestic house construction in America and Europe today.

So, while wood use in large construction virtually ceased from the late nineteenth century, it continued as an inherent material in small-scale construction. At the same time, the overall demand for wood continued, but not for the same type of wood that was traditionally used in shipbuilding or construction. From the middle of the nineteenth century, scientists and researchers were already exploring new uses for wood and wood fibre that would dramatically increase its use in supplying new markets. These markets would also require not only different approaches to engineering wood, but also different tree species. Scarce European hardwoods such as oak and beech were replaced by softwoods such as spruce and pine as a new timber market dynamic evolved.

Softwoods provided new opportunities to develop a range of products whereby all the tree could be used commercially. Instead of using small lengths just for firewood, research results in the nineteenth century showed that small wood, including waste wood, could also be converted into products unrecognisable from their original state and turned into a range of paper and board products. In the case of paper conversion, this wasn't a new development as the Chinese had produced paper from the mulberry tree since the sixth century, but not on an

HARDWOOD VERSUS SOFTWOOD

In the northern hemisphere, hardwood is the generally accepted term for the timber of broadleaf, mainly deciduous trees as opposed to softwood, which refers to coniferous, mainly evergreen trees. The terms relate to the botanical grouping of trees rather than their hardness or softness as some conifers such as yew (above) are harder than many broadleaves including poplar and lime (below), but generally broadleaves are harder and have a higher wood density than conifers.

Hardwoods are used for furniture, joinery, veneering and cladding while softwoods are widely used as structural timber both in solid and engineered form. The volume of coniferous sawn and traded internationally is approximately five times greater than the volume of non-coniferous wood. Tropical hardwoods are mainly evergreen and depending on the region often show continuous growth throughout the year so they don't have obvious annual growth rings. However, where there are seasonal temperature variations, annual rings are defined clearly.

industrial scale as in Europe. While hardwoods can be pulped for paper, softwoods are far more suitable and efficient. Unlike hardwoods, they are plentiful, grow straighter and produce high yields even on poor soils, in exposed conditions. Despite increased digitisation and a shrinking newsprint market in recent years, demand for paper and paperboard is forecast to increase to 482 million tonnes by 2030, according to Timo Suhonen and Tomin Amberla in their study for Pöyry Management Consulting in 2015. In addition to producing paper products from small – and waste – wood, softwood timber is being reprocessed to produce wood-based panel products while the larger lengths are being used not only as sawn timber in housing but also engineered for strength and adaptability for high-rise buildings. There is no waste in wood as every part of a softwood tree, including its bark, branches and even foliage, can be utilised. The aim is to achieve a balanced softwood and hardwood mix, especially in a country like Ireland which has lost most of its native hardwoods over the millennia.

Getting the balance right between the restoration of the hardwood resource and the continuation of a viable softwood resource is a major challenge facing Irish forestry. Hardwoods will continue to play a major role in furniture, turnery, panelling and small-scale building but where functionality, flexibility and scale are required, softwoods – especially engineered softwoods – are the species of choice for large-scale projects in construction, energy and new revolutionary applications in the bioeconomy as well as renewable energy in district heating, public buildings and industry. Increased use of softwoods should not be at the expense of hardwoods. Rather, sustainable use will see both working side by side. In Ireland, where hardwoods are scarce, increased use of commercial softwood takes the pressure off harvesting slow-growing hardwoods, especially native species. While markets may influence species choice, ultimately soil, will dictate species productivity. So, when establishing new plantations or regenerating existing forests, it is important that forest owners not only match the species to the market but also to the underlying soils and site conditions that will dictate their quality and productivity. And there are significant differences between the requirements of softwoods (Figure 6.1) and hardwoods (Figure 6.2). That way, maximising their ability to sequester carbon in the forest which is a key answer to "Why Forests?" is counterbalanced by their ability to store carbon in their after-forest life use, which is main answer to the question "Why Wood?".

This extension of the life of a tree in carbon storage and in displacing fossil-based material is essential in understanding the role for the forest and wood bioeconomy, especially in climate change mitigation.

As outlined in Chapters 5 and 6, getting the balance right between what's growing in the forest and what's removed is the key to sustainable supply and the important role wood plays in climate change mitigation.

TYPICAL TIMBER USES FOR CONIFERS (SOFTWOODS) SUCH AS SPRUCE AND

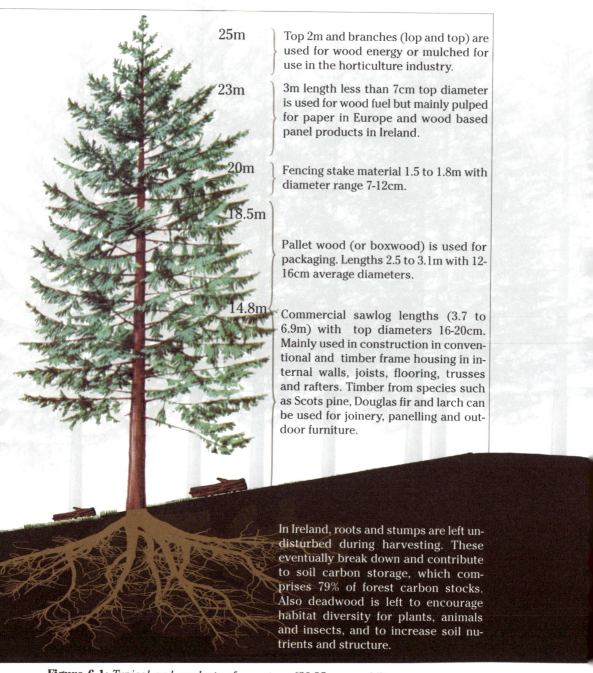

25m — Top 2m and branches (lop and top) are used for wood energy or mulched for use in the horticulture industry.

23m — 3m length less than 7cm top diameter is used for wood fuel but mainly pulped for paper in Europe and wood based panel products in Ireland.

20m — Fencing stake material 1.5 to 1.8m with diameter range 7-12cm.

18.5m — Pallet wood (or boxwood) is used for packaging. Lengths 2.5 to 3.1m with 12-16cm average diameters.

14.8m — Commercial sawlog lengths (3.7 to 6.9m) with top diameters 16-20cm. Mainly used in construction in conventional and timber frame housing in internal walls, joists, flooring, trusses and rafters. Timber from species such as Scots pine, Douglas fir and larch can be used for joinery, panelling and outdoor furniture.

In Ireland, roots and stumps are left undisturbed during harvesting. These eventually break down and contribute to soil carbon storage, which comprises 79% of forest carbon stocks. Also deadwood is left to encourage habitat diversity for plants, animals and insects, and to increase soil nutrients and structure.

Figure 6.1: *Typical end products of a mature (30-35 years old) coniferous (softwood) tree such as spruce with a height of 25m and a volume of approximately 1.5m³. Unlike broadleaves, conifers have straight stems from base to tree top, so a much higher percentage of each tree is available for sawn commercial construction use from as early as 25 years of age. Small logs can be converted to stake wood, packaging and reprocessed into wood based panel products from year 15.*

BROADLEAVES (HARDWOODS) SUCH AS OAK

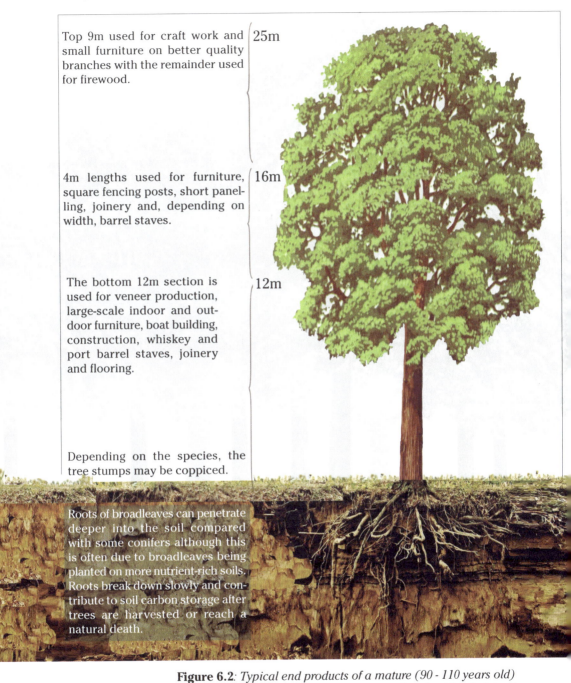

Top 9m used for craft work and small furniture on better quality branches with the remainder used for firewood.

25m

4m lengths used for furniture, square fencing posts, short panelling, joinery and, depending on width, barrel staves.

16m

The bottom 12m section is used for veneer production, large-scale indoor and outdoor furniture, boat building, construction, whiskey and port barrel staves, joinery and flooring.

12m

Depending on the species, the tree stumps may be coppiced.

Roots of broadleaves can penetrate deeper into the soil compared with some conifers although this is often due to broadleaves being planted on more nutrient-rich soils. Roots break down slowly and contribute to soil carbon storage after trees are harvested or reach a natural death.

Figure 6.2: *Typical end products of a mature (90 - 110 years old) broadleaf (hardwood) tree such as oak or beech with a height of 25m and a volume of approximately 1.5m³. Compared to conifers, broadleaves have a smaller proportion of straight stems. However, the commercial proportion has high added value uses in furniture, joinery, flooring, veneering and panelling for some species. These generally need growing rotations in excess of 100 years to achieve yields similar to commercial conifers.*

SUSTAINABLE WOOD SUPPLY

Continuity of supply is critical for increased use of wood as a long-term alternative to fossil-based materials. This requires an appreciation of the long-term nature of forestry. Most countries that overcut their forests face major problems in correcting deforestation. Ireland and Britain, like other European countries, experienced deforestation from the seventeenth century.

Europe responded in time to restore its forest cover especially from the nineteenth century. Ireland and Britain didn't respond until the early twentieth century, which is why both islands still haven't developed forest and wood cultures. Forest – and timber – depletion was observed by poets such as Aogán Ó Rathaille (c.1670-1726) in Ireland and Andrew Marvell (1621-78) in England. Ó Rathaille attributes the decline of Irish forests to the coloniser ("The English bleater") while Marvell points the finger at architects in his poem, "Dialogue between the Soul and the Body":

> *So architects do square and hew*
> *Green trees that in the forest grew.*

Almost four centuries have passed since architects, along with their teams of carpenters and craftsmen, squared and hewed timber with such abandon that led Marvell to question the relationship between forests and their end uses – and users – in what is one of the earliest poems on sustainability, believed to be written in 1652 and published posthumously in 1681. Marvell drops in these final lines which, according to John Carey are presented as "A teasing, intriguing doubleness …". (Marvell liked to have it both ways in some of his poems especially in "An Horatian Ode Upon Cromwell's Return From Ireland".) Carey asks: "But are trees superior to architecture, or architecture to trees?" He speculates that Marvell "does not decide", but Marvell may well have decided. He would have been aware of the over-exploitation of forests during his lifetime for ship building, construction and charcoal-burning for smelting, and may well be deliberately sowing the seeds of doubt – and guilt – among construction and naval architects. Today, he might sow the seeds of guilt among architects, engineers, designers and builders not for their overexploitation of trees but for their over-reliance on fossil-based building materials rather than wood.

What Marvell saw in England – like Aogán Ó Rathaille, in Ireland – was the depletion of a once-great woodland resource. While Ó Rathaille lays the blame at the door of the coloniser, deforestation also occurred in England. There seems to be little distinction between the coloniser and the colonised when it came to forest destruction. Even today, forest cover, at 10% of the land area in England, is less than Ireland. The UK's lamentable effort at restoring its forest resource is rescued by Scotland where forest cover is a commendable 19%.

SURPLUS WOOD COULD HOUSE ALMOST TOTAL EU POPULATION

Dr. Pablo van der Lugt (below), author and lecturer in bio-based building at Delft University, advocates a major timber house and apartment construction programme for Europe. He outlines how timber is now gaining momentum as the medium of choice not only in small-scale buildings but also in medium to high rise construction. At its simplest, his premise is "One tonne of softwood replaces one tonne of concrete or steel and saves 1.5 tonnes of CO_2."

In *Tomorrow's Timber: Towards the Next Timber Revolution* – co-authored with Atto Harsta – he maintains that "almost the total EU population increase could be housed in mass timber in the future". This scale of wood construction would place unprecedented demand on wood supply but he argues that there is sufficient timber in European forests to satisfy this market because Europe is not optimising timber harvesting. "The annual increment in Europe's forests is 800 million m³, while we harvest 500 million m³," he says. "Apart from Sweden no other country harvests its increment." Dr. van der Lugt is not advocating harvesting all the 300 million m³ balance as some of it may be inaccessible or not of the desired quality, but he said "we could build a half million houses from the 70 million m³ increment that is not being harvested at the moment".

While England afforests mainly native species for biodiversity reasons, Scotland established an average mix of 8,000ha of productive conifers and 3,500ha of mainly native broadleaves to ensure both wood and non-wood functions from 2018 to 2021. Scotland's native afforestation programme is larger than the total planting programme of the rest of the UK combined.

Stuart Goodall, director of Confor, believes that the UK, which is the second-largest importer of timber in the world, has an economic, environmental and moral obligation to produce its own timber. Unlike past demand for wood, which was sourced in native natural woodlands, future demand will concentrate mainly on man-made forests, leaving the native woodlands either protected or harvested sparingly. Getting this balance right is critical to ensure future supply, especially if wood use is increased significantly to displace fossil-based materials in construction, energy, packaging and conversion to fuel.

Chadwick D. Oliver and fellow authors maintain that there are more than sufficient timber reserves. They calculate the annual global harvest at 3.4 billion m^3. This is equivalent to one fifth of annual global wood growth which is estimated at 17 billion m^3. They contend that increasing the wood harvest to the equivalent of 34% or more of annual growth would ensure:

> Between 14% and 31% of global CO_2 emissions could be avoided by preventing emissions related to steel and concrete; by storing CO_2 in the cellulose and lignin of wood products; and other factors.

> Between 12% and 19% of annual global fossil fuel consumption would be avoided if scrap wood and unsaleable materials were burned to produce energy.

> Wood-based construction consumes much less energy than concrete or steel construction. Through efficient harvesting and product use, additional CO_2 is saved through the unreleased emissions, materials, and wood energy than is lost from the harvested forest.

Oliver makes the case for greater timber use in its own right but specifically to reduce dependence on fossil-based materials especially the manufacture of steel, concrete, and brick which accounts for about 16% of global fossil fuel consumption. They claim that when the transport and assembly of steel, concrete, and brick products are included, their share of fossil fuel burning is closer to between 20% and 30%. If wood becomes a central material in climate change mitigation, it will increasingly replace carbon-intensive steel and concrete. As a result, Gresham House forecasts 5.8 billion m^3 of timber will be required by 2050. This forecast rise in timber demand is set against a constrained supply which is limited by the long-term growth rate of the trees and limited land area. Current annual forecasts for long-term, sustainable timber supply between now and 2050 range between 3.7 billion m^3 and 4.7 billion m^3. If, as Oliver maintains, we are only harvesting 20% of annual timber increment and with future increased afforestation, the global forest is capable of meeting future demand. There are caveats in realising future timber production forecasts. Deforestation is still a major global issue and even in regions where sustainable forest management is practised, future timber reserves may be eroded by increased forest damage such as fire, insect and fungal attack, drought and windthrow, which are all linked to global waming.

WOOD AND CLIMATE CHANGE

Forests and wood are inextricably linked with climate change. Deforestation and over-reliance on fossil-based construction material and fuel, since the Industrial Revolution, marked the beginning of humankind's major negative impact on climate change and the resultant increase in greenhouse gas emissions. Excessive demand for illegally felled tropical hardwoods – especially from the northern hemisphere – combined with unsustainable use of fuel wood and clearances for agriculture are proving catastrophic not only for these deforested countries, but also globally.

We simply cannot make the case for wood as a sustainable living resource as long as illegal timber trade and deforestation are tolerated. These twin global environmental disasters, which once threatened forestry in developed countries, have almost destroyed forest development in some tropical and sub-tropical countries. Up until recently, they were tacitly encouraged by developed countries, which have played a major role in exploiting the ever-decreasing wood reserves and their ecosystems in these regions.

This has been acknowledged since the first Earth Summit at Rio in 1992 when conservation and protection of forests were major topics. The debate on the future role of forests in climate change mitigation also included sustainable timber trade. Delegates from 130 nations agreed a broad statement of principles for protecting forests and also addressed illegal logging and sustainable timber trade in the International Tropical Timber Agreement.

Forestry and its role in achieving net-zero greenhouse gas emissions by 2050 has remained central to climate change summits since Rio. This was highlighted in the Paris summit and upheld in Glasgow's COP 26 in 2021. The 2021 report stated: "Unless there are immediate, rapid and large-scale reductions in greenhouse gas emissions, limiting warming to close to 1.5°C or even 2°C will be beyond reach."

The role that forests play in sequestering CO_2 from the atmosphere is undisputed. It is acknowledged in the EU Forest Strategy and Ireland's Climate Action Plan, which maintains that afforestation is Ireland's "single largest land-based climate change mitigation measure". Equally important, the Plan, also acknowledges the role of wood and wood products as carbon sinks in displacing fossil-based material:

Ireland's Climate Action Plan 2021 and successive plans have identified afforestation as the "single largest land-based climate change mitigation measure" and wood for construction as a "substitute for conventional carbon-heavy construction products, such as concrete, brick and steel".

> Using wood and wood-based products for construction is a sustainable substitute for conventional carbon-heavy construction products, such as concrete, brick and steel.

Promoting the dual benefits of forestry – carbon sequestration in the forest and carbon storage in wood use – is central to the climate change debate. Failure to acknowledge this duality results in a disconnect between the forest and the product and between the forest owner and those who use wood including processors, manufacturers, architects, engineers, designers, wood workers, and restorers as well as wood scientists and researchers who wish to take wood to another level in its role to decarbonise the economy. Wood has two major roles in climate change: storing carbon in timber products, and displacing fossil-based materials, especially in construction.

BEYOND WOOD

Today's scientists and researchers are going beyond wood's physical traditional applications as they breakdown wood fibres into minuscule fibrils and restructure them into new biomaterials. Now, wood – and wood waste – is transformed as it is reprocessed into biofuels, fertilisers, textiles and in new forms of packaging to displace plastic.

This approach emboldens Karl-Henrik Sundström, former CEO of the Swedish-based forestry company, Stora, to make the claim that "Everything that's made with fossil fuels today can be made from a tree tomorrow". This is more than a brave – even audacious – futuristic boast as it is rooted in an innovative wood tradition. Rather, it is a continuation of generational experimentation, from turpentine extraction from North American and European pine species to wood transformation to paper in the nineteenth century and more recently into engineered wood products. Sundström believes wood can be taken to another level as outlined in an interview with Jan Lindroth. He explains that the forest industry is now seen as an important sector in "breaking away from fossil fuel dependency". Sundström believes the answer lies in the tree:

> If we are to meet the Paris climate goals, the use of fossil-based materials must be quickly reduced and replaced with renewable materials. Materials made from trees, such as paper and board, can be reused five to seven times, and trees can be grown again. That is one reason why our industry has been in the spotlight…. The new role of the tree is two-fold: firstly, its cellulose is being used for products beyond paper, board and tissue, and secondly, all parts of the tree are being utilized …

Sundström looks beyond the physical visible characteristics of wood to explore its main constituents – lignin, cellulose, hemicellulose, and extractives, as he discusses their possibilities with Lindroth:

> Today, lignin and hemicellulose are largely burnt for energy ... but Stora Enso is exploring many new applications, some of which are already on the market. Lignin is used to make resins and adhesives used in, for example, construction, but it is also possible to make carbon fibre from lignin, which could make it an alternative to steel. Hemicellulose is used in applications such as sweeteners. Additionally, it has great potential as a raw material in a wide range of chemicals. Cellulose is used in composites that can be moulded into all kinds of shapes, and nanocellulose, a derivative of cellulose, is used to make lighter packaging, thinner film and more, as well as replacing plastics in many areas.

He acknowledges the global challenge in reducing single use plastics. Annual global plastic production now averages more than 300 million tonnes, compared with 1.5 million tonnes in 1950, according to Ian Tiseo. Within 70 years, more than half of the nine billion tonnes of plastic materials produced on our planet have ended up as waste which will take up to 500 years to decompose. Hannah Ritchie and Max Roser illustrate the enormity of the challenge in reducing mismanaged plastic waste that is discharged into the world's oceans. They estimate that 80% of ocean plastics derive from land-based sources, with the remainder coming from marine sources.

Sundström maintains that likely bans on single-use plastics are forcing retail and the hospitality sector to quickly find renewable, recyclable and fossil-free materials and products. Responding to this challenge, in 2019 Stora Enso Oyj launched Cupforma Natura Solo, a new renewable paperboard suitable for hot and cold drinking cups which the company claims "is produced without a traditional plastic coating layer and designed for full fibre recovery in a recycling process". In Canada, research is underway "to scale up production of a new lightweight wood-fibre-based composite material to create automotive parts", which according to Natural Resources Canada (NRC):

> [W]ill combine wood pulp with polymers to create a special strong and lightweight thermoplastic with more uniform and improved properties compared to other products. The new product, if successfully applied in the automotive sector, could have a number of consumer and commercial applications, including aerospace parts, pharmaceuticals, solar panels and cosmetics.

Wood is the basis for the manufacture of a variety of common household products as illustrated by NRC. Some paints contain the wood component hydroxyethyl cellulose (HEC), a gelling and thickening agent used to match the texture of a liquid product to consumers' needs. HEC is also widely used in cosmetics, adhesives, detergents and other household products. Even some bath towels are made with rayon, which is produced from the wood component cellulose.

Moving away from fossil fuels to renewable cellulose-based liquid fuel is becoming a reality for automobile and aviation companies. Cellulose from wood is produced in larger quantities and at a faster rate than anything else on the planet so there is a renewable supply of biomass. Nordic countries and companies are taking the lead. Södra has invested in the production of biomethanol, a sustainable fuel from forest raw material. Similar research is being carried out in Canada, while in the US Forisk Consulting has teamed with the Schiamberg Group to evaluate US-based projects and technology pathways in producing biofuels.

Cupforma Natura Solo by Stora Enso is a renewable material which displaces plastic when used to make paper cups that are suitable for hot and cold drinks.

Apart from producing biofuel, wood is also likely to play other roles in the automotive industry. Wood, or the conversion of lignin in wood to hard carbon anode materials for lithium-ion batteries, can play a major role in replacing the fossil-based anodes, mainly synthetic graphite, currently used in batteries.

Stora Enso demonstrates that wood-based carbon can be utilised as a crucial component in batteries used in the automotive industry as well as consumer electronics, and large-scale energy storage systems.

Wormslev, and fellow researchers, outlined the potential and challenges in their report "Sustainable Jet Fuel for Aviation", commissioned by the Nordic Council of Ministers. They cautioned that even in the climate change conscious Nordic countries, implementing commercial production of sustainable jet fuel will not be achieved without cohesive political support including the promotion of public-private partnership "between the aviation sector, jet fuel producers, universities and other public entities". Overall, the authors are positive:

> The conclusions of the study are that Nordic production of sustainable biojet fuels has potential benefits stemming from the reduction of GHG emissions and negative environmental impacts, the development of new technology, as well as economic growth and job creation.

Breaking wood down into its component parts to displace fossil based materials can also benefit agriculture by producing biochar. This is achieved by decomposing wood at high temperature, in a process called slow pyrolysis. While bio-based fuel is produced during pyrolysis, biochar enhances agricultural production. It can be returned to the soil where it absorbs carbon and stores it for years while also improving the soil's condition by improving water retention and increasing nutrient levels. In addition to locking in carbon, it can also help decrease methane and nitrous oxide emissions from the soil.

Global research demonstrates that wood is finding new renewable uses in displacing carbon-based plastics and fuel as well as fertilisers. But it is in construction that it is reclaiming lost ground, not only for its inherent strength and aesthetic characteristics, but also because it is capable of replacing the very same fossil-based materials that supplanted it in the first instance.

Wood acts as a carbon store in the forest and mirrors this role as a carbon sink after it leaves the forest. While these undoubted climate change benefits will increase the drive to use more wood, it has to be a medium of choice based on its inherent strength, aesthetic, acoustics, durability and design qualities. It has no divine right as a medium of choice despite obvious advantages over non-renewables. It has to withstand structural and economic stress testing like all materials, so that architects, engineers and designers have confidence in wood in order to specify it for a wide range of uses either in sawn, engineered or other iterations.

BUILDING IN WOOD – LOW, MEDIUM AND HIGH RISE

While the potential of wood in displacing fossil fuel energy and plastics is an exciting twenty-first century development, this revolution is well underway in construction. Wood's future in housing and low-rise building is assured as timber frame is well-established in the construction market globally as outlined by BM Trada:

> Over 70% of the population of the developed world live in timber frame housing. In Canada and the US, over 90% of low-rise buildings use timber frame technology. In Scotland, three quarters of all new build houses and up to 90% of UK self-build houses are timber frame. Mainstream house builders are also recognising the benefits in speed of build which prefabricated timber building systems can offer.

The trend of increasing market share for timber frame buildings is likely to continue, not only because of its climate mitigation benefits, but also because timber frame construction is fast and safe. Irish builders and architects have been slow to adopt timber frame construction but this is changing (see Wood Innovation in Ireland, page 144).

REDUCING EMISSIONS: WOOD VERSUS FOSSIL – BASED MATERIALS

During its production, one tonne of:
- Concrete releases 159kg of CO_2 into the atmosphere.
- Steel releases 1,240kg of CO_2 into the atmosphere.
- Aluminium releases 9,300kg of CO_2 into the atmosphere.

In contrast wood absorbs a net 1,700kg of CO_2 from the atmosphere, over and above the energy expended in growing, harvesting and processing.

Source: *New Zealand Forestry Owners Association – Forestry Facts & Figures, 2018.*

ENGINEERED OR MASS WOOD?

Engineered wood includes a range of timber products which are manufactured by binding boards of varying lengths together with adhesives or other methods such as dowels or nails. Also known as "mass timber", engineered wood is the primary load-bearing structure in small and large-scale buildings. It can replace or complement steel and concrete as a structural solution, which is why it is highly regarded in the role it plays in climate change mitigation.

The most commonly used species are softwoods such as Norway spruce, Scots pine, Douglas fir – used in Brock Commons Building, Vancouver (pictured) and larch. Softwoods are preferred because they are durable, fast-growing, straight and readily available.

Hardwoods such as oak, sweet chestnut and ash are also suitable but lack the straight form, yield and availability of softwood logs. Instead, these are engineered for high value interior use such as furniture, which provide opportunities for producing a wide range of designs that are not limited to straight lines.

Wood has been used in reconstituted form in wood-based panels (WBPs) for over a century, while engineered or "mass wood" is now regarded as a major construction medium.

Engineered wood has revolutionised the construction industry because, unlike timber frame, it forms the structural element of medium-to-high-rise buildings and substitutes for most of the main fossil-based materials. Reducing and displacing non-renewable fossil-based materials is a major challenge in the global construction sector. The built environment generates nearly 50% of annual global CO_2 emissions according to *Architecture 2030*, which adds:

> Of those total emissions, building operations are responsible for 27% annually, while building materials and construction (typically referred to as embodied carbon) are responsible for an additional 20% annually.
> [...]
> Just three materials – concrete, steel, and aluminum – are responsible for 23% of total global emissions (most of this is used in the built environment).

Because engineered wood is used in the structural element of building, it has a major role to play in reducing CO_2 emissions. The most exciting development in twenty-first century timber construction lies in the role engineered wood will play in large-scale buildings. Architects have been building tall in wood for centuries. Some of these buildings, especially in China and Japan have even outlasted stone. For example, the 67-m Pagoda of Fogong Temple, in Shanxi province, China, built in 1056, is made entirely from wood, apart from the stone base. It is still a functioning building having survived extreme weather conditions and numerous earthquakes over the centuries. Like other long-lasting buildings, it is an inspiration to modern architects as it highlights wood as a structurally sound, durable and aesthetically pleasing material as well as a sustainable and renewable medium. It also demonstrates wood's longevity as the species used for the temple was Xingan larch, sourced in the forests of Northern China almost a 1,000 years ago.

Wood specifiers including architects, engineers and designers who wish to use wood in large-scale projects can no longer source Xinguan larch and other scarce species in sufficient quantities to build high. Even if they could, they would still be faced with challenges in specifying wood because of its variability. Wood properties vary from species to species, from tree to tree, and even within a tree, so variability in wood is challenging for architects and engineers who are used to working with steel and concrete, especially for large-scale projects. While variability has aesthetic benefits, it is a disadvantage when uniformity and accuracy are required in construction.

Engineering wood solves the problem of variability. It allows architects to design high-rise buildings in mass timber including glulam (GLT), laminated veneer lumber (LVL) and cross-laminated timber (CLT). These have the potential to

The 67-m tall Fogong Temple in Northern China has survived more than 1,000 years of storms and earthquakes since it was built in 1056.

transform the construction industry to a low carbon-built environment. Since the 50-m Brock Commons building in Vancouver was built in 2016, a number of buildings have reached 80m tall, with plans to surpass this height in North America, Europe, Japan and China.

Mass timber is creating a revolution in large-scale timber construction and this will continue now that regulations and standards are in place that satisfy planners, builders and specifiers that wood conforms to fire regulations. It is therefore important to discuss the buildings that have created these standards while convincing private and public decision-makers of the value of wood in sustainable construction. These include Brock Commons, Tallwood House in Vancouver Canada, and Treet (The Tree) in Bergen, Norway.

Brock Commons Tallwood House began the revolution with an innovative 18-storey tall wood hybrid building at the

MASS WOOD PRODUCTS

Although, the number of engineered wood product is increasing, the most commonly used in Europe are cross-laminated timber (CLT) and glue-laminated timber (GLT) or glulam often referred to as mass wood.

CLT is a thick timber panel product manufactured by gluing together several layers of timber boards, with successive layers glued at right angles (pictured above). These can have final dimensions up to 0.4m thick, 2.9m wide and 18m long and can be used for building floors and walls. CLT provides strength, rigidity, dimensional stability and high load-carrying capacity for building up to 80m tall while there are plans to build higher.

Glulam is manufactured from layers of parallel timber laminations. Individual laminates can be finger-jointed to produce long lengths. Glulam can be manufactured in a wide variety of shapes, sizes and arrangements including straight and curved beams up to an individual component length of 50m. Fine examples of glulam in Ireland include Beyond The Trees, Avondale and Cherrywood Timber Canopy, Dublin (pictured below).

University of British Columbia (UBC). The wood structure was completed less than 70 days after the prefabricated components arrived on site, approximately four months faster than a typical project of this size. Brock Commons is the student building at the University of British Columbia (UBC). The mass timber building is built over a concrete podium and has two concrete stair cores. The floor structure consists of CLT supported on glulam columns. The roof is made of prefabricated sections of steel beams and metal decking. The total CO_2 equivalent avoided by using wood products over other materials in the building is more than 2,432 tonnes, as estimated by Sathre and O'Connor using the Wood Carbon Calculator for Buildings.

Engineered wood is also used in Treet (The Tree), a 49-m tall building with a floor area of 4,500m². Completed in December 2015, it was designed as a multipurpose 14-storey building using two forms of mass timber "in a hybrid combination of a glulam timber and CLT". While 85% of the building comprises wood, ARTEC understands, and acknowledges the importance of other materials to ensure stability and protection against weather:

> Powerfloors consisting of one-storey-high trusses, combined with concrete slabs on two levels, give the structure enough strength and weight to remain stable in the wind. Treet is clad in Corten steel to the east and west, and glass to the north and south, to protect the timber from the wet weather conditions in Bergen.

Treet includes 62 apartments, a fitness room and a rooftop terrace. The building has a glulam timber framework, while CLT was used for stairs, elevator shafts and balconies. The building satisfies Norway's strict fire regulations. Before Brock Commons and Treet were completed, two buildings were already taking shape to reach heights in excess of 80m.

Austrian architecture practice Rüdiger Lainer and Partner (RLP) designed the 84-m HoHo wooden skyscraper in Vienna which was completed in 2019. Over 75% of the 24-storey structure is in wood, which saves 2,800 tonnes of CO_2 emissions over similar structures built in steel and concrete, according to RLP. In addition, all the wood is sourced from local Norway spruce, cutting the cost of delivery from forest to CLT sawmill and prefabrication plant before eventual transport to the HoHo building site.

Further north in Brumunddal, Norway, the Mjøstårnet, also completed in 2019, surpassed the HoHo in height by less than half a metre. The 85.4-m-tall Mjøstårnet (the tower of lake Mjøsa) was the world's tallest timber building by the end of 2020. Designed by Voll Arkitekter, this multiuse building consists of glulam trusses, columns, and beams, while CLT was used for stiffening elements, and to build the elevator shafts and the staircases.

As mentioned, plans are already in place to exceed 80m in global multipurpose buildings. The most ambitious project is planned by Japanese wood products company Sumitomo Forestry. It plans to build a 70-storey, 350-m-tall hybrid mass timber skyscraper by 2041. Designed by Nikken Sekkei, it will celebrate the 350th anniversary of Sumitomo Forestry, nine years before the world should achieve net zero.

WOODNOTE: THE ARCHITECTURE OF SOUND

Reinventing the role of mass wood in contemporary construction is rooted in past achievements especially in countries that have strong wood cultures. This revival of wood as a building material will face opposition from more powerful voices that advocate business as usual concrete, steel and other non-renewable fossil based materials. Despite changing fashions and fortunes, wood's aesthetics provides it with a clear advantage over its fossil based rivals. Our encounter with wood is an aesthetic experience that is sensory in its broadest meaning as it provides a connectedness that involves touch, sight and smell. The physical experience also extends to sound where wood's aesthetic is matched by its acoustics.

Architects, designers, musicians and librarians have long recognised the advantages of wood as a medium that can enhance musical performance in concert halls and opera houses or create an atmosphere of serenity and stillness in libraries.

Depending on the design and tree species, wood can

The 85.4-m-tall Mjøstårnet, Norway was the world's tallest timber building by the end of 2020.

absorb or reverberate sound waves. It has retained its presence in the world's great libraries because its aesthetics are compatible with its acoustics, in particular its ability to absorb sound waves, which is essential where an atmosphere of calmness is required. These requirements date to the Renaissance, when Michelangelo acknowledged its place when designing the Laurentian Library, Florence. For all its beauty, the grey stone staircase was not Michelangelo's material of choice. In correspondence with Bartolomeo Ammannati (who completed the staircase in 1559) he was deeply unhappy that wood had not been used. "His preferred material" for the staircase, according to Michael Levey, "was not stone but wood – walnut – which he felt would be more in keeping with the library, with its wooden desks and wood ceiling".

Wood is impossible to avoid in library design from the great European libraries constructed from the sixteenth century right up to contemporary structures. All the elements that constitute a relaxing educational environment can be made from wood including tables, chairs, book shelves, panelling and flooring and to varying degrees, in construction. Even the very books that line the shelves from the late nineteenth century have been made from trees when pulpwood was converted to paper on an industrial scale, replacing the previous costly linen, fabric and animal vellum-based paper.

It is doubtful if the Long Room Library, Trinity College Dublin would be such a tourist magnet if it retained the plaster ceiling of the original building, constructed between 1712 and 1732. Undoubtedly, it would still be a major attraction for academics and researchers, but the decision to raise the ceiling in the nineteenth century to add further space for the 200,000 oldest books was a defining statement for the library and the college. The 65-m long barrel-vaulted oak ceiling replacement, along with an extra upper gallery row of wooden bookcases, was an inspired structural and aesthetic choice. Completed in 1861, the wood-dominated design ensured the Long Room Library as an essential destination for scholars and tourists alike. The tradition of wood use and design continues in contemporary international libraries.

The 1987 Wallach Division of the New York Public Library is an excellent example, while the Calgary Central Library and the Central Library Oodi in Helsinki – both completed in 2018 – use wood, sourced from their heavily forested hinterlands, extensively.

In Ireland, wood use finds expression in two Co. Dublin libraries – dlr Lexlcon, Dún Laoghaire, completed in 2014, and Ballyroan in 2012. While the exterior design of dlr Lexlcon provcd controversial, the interior received unanimous approval because of its extensive use of oak in the bookshelves and panels for acoustic modulation, and in the staircase and floors. In Ballyroan, birch and oak are used in the gallery space and ceilings in a building which combines library facilities with separate spaces for events and exhibitions.

Ballyroan Library, Co. Dublin, completed in 2001, features birch and American white oak throughout the interior.

Long Room Library, Trinity College Dublin featuring the oak 65-m long barrel-vaulted ceiling and the upper gallery bookcases completed in 1861.

While wood has all the absorption characteristics to provide stillness in library design, a different set of acoustic qualities is required for opera and concert halls. Wood is an obvious choice in this creative environment when it becomes an extension of the musical instrument, as wood provides the material for all the orchestral string section and was the original material for making woodwind instruments before they incorporated metal. For centuries, wood has been regarded as a unique material for making musical instruments, with different species providing their own unique sound, most celebrated by the Cremonese violin makers Antonio Stradivari (1644-1737) and Giuseppe Guarneri (1698-1744). They chose ebony for the fingerboard and European maple for the back and ribs while the tonewood was always Norway spruce sourced from the oldest trees in upland cold regions. A multiplicity of woods is used in making classic grand pianos, including spruce, hornbeam, beech, mahogany, larch, boxwood, maple, poplar, walnut, baya and ebony.

It is no coincidence therefore that innovative architects opt for wood, especially for concert hall interiors. Wood finds extensive articulation in amplifying and absorbing sound waves to enhance all music: from the solo singing voice to choral and orchestral music. But architects can't do it on their own. Maximising the acoustics of concert halls requires a deep scientific and technological knowledge of different species and their positioning which is acknowledged by the architect Renzo Piano.

He worked with acoustic consultant Helmut Muller in exploring the use of lasers to trace reflections of sound waves when designing the three concert halls of Parco della Musica in Rome. Like musical instruments, there is a wide range of wood species at the architect's disposal in designing concert halls. Piano opted for cherry, sourced from an American Hardwood Export Council (AHEC) supplier in eastern USA, not just because of its aesthetic appeal but because it also satisfies demanding acoustic and structural criteria.

Piano recognised the importance of creative architecture and music in the lives of the people of L'Aquila city, Italy when their concert hall was damaged by the devastating earthquake in 2009. Within three years, Piano and his team had created three wooden buildings or cubes including a 238-seat concert hall. This temporary flat-pack building used vividly painted European larch in the exterior which brings colour, music and hope back to the people of L'Aquila until their original building is restored.

In the Tokyo Opera City Concert Hall, opened in 1997, principal architect Takahiko Yanagisawa opted for ridged oak to optimise the acoustic conditions and the aesthetic brilliance of the hall culminating in the spectacular pyramidal ceiling. Like Piano, Yanagisawa hired an acoustic expert to advise on the positioning of the solid oak which, depending on the angles of the grooves, can diffract and reflect sound. He chose audio pioneer Leo Branek, who provided the science to achieve the

Opposite: Renzo Piano used cherry in the interior of Parco della Musica, Rome both for its aesthetic and acoustic qualities.

optimal sound balance and performance.

The Sibelius Hall in Lahti in southern Finland, completed in 2000, takes wood a step further as it is used in the main structure as well as the interior. This homage to wood is also a tribute to Sibelius, who was influenced by nature and the forest, in particular his tone poem *Tapiola*, which portrays Tapio the forest spirit. The architects Hannu Tikka and Kimmo Lintula specified birch, Norway spruce and Scots pine, which are the main species in Finland's forests. Oil- and heat-treated birch is used for the floor of the main hall while softwood glulam is used for the load-bearing structure of the building. Wood was the natural choice for the Sibelius Hall, not least because of the abundance of timber close to Lahti – forests cover 75% of the land area of Finland.

Although Ireland lacks a wood culture, Keith Williams Architects and the Irish Government's Office of Public Works Architects Department decided on a wood-dominated interior for Wexford Opera House but they had to travel to find a suitable species. Eventually, black walnut was chosen, sourced from sustainably managed forests in eastern Canada. Matt O'Connor, who produced the initial feasibility study for the opera house, said: "Its rich dark brown colour similar to the appearance of wood used in stringed musical instruments, added enormously to the rich aesthetic of the auditorium."

L'Aquila temporary city concert hall, used painted European larch panels on the exterior. The building was completed within three years of the devastating earthquake which destroyed the original concert hall in 2009.

WOOD INNOVATION IN IRELAND

When a country loses its forests and woodlands, it is inevitable that their disappearance will be accompanied by a loss of wood culture. While Ireland is not the only country in Europe that has witnessed forest decline, it is unique in its slowness to arrest this decline. Significant forest restoration only began in the middle of the last century. This was carried out by the State to reduce Ireland's dependence on imported timber.

Sibelius Hall, Lahti, Finland. Wood was used extensively in the concert hall interior and exterior.

As a result, Ireland developed a masonry culture in construction, which is now beginning to change as we are self-sufficient in softwoods due to a major afforestation drive over the past 70 years. But the full potential of wood has yet to be realised in public and private construction projects, unlike our European neighbours who have an innate confidence in timber as a reliable and environmentally friendly resource.

While the gap between wood use in construction is narrowing in Europe and Ireland, it remains vast in sectors such as joinery, furniture manufacture and other domestic wood applications that require hardwoods, although some softwoods can be used in these specialist markets. While Ireland has established a major softwood resource, because of the poor land made available for afforestation, the country has a dearth of hardwoods. Although broadleaf afforestation has increased from 5% of all planting in the twentieth century to over 30% since 2000, most of these forests won't mature until the end of this century while our most common hardwood – ash – has succumbed to the ash dieback disease. We don't have a commercial indigenous hardwood manufacturing industry to cater for the furniture, panelling, joinery and flooring markets. We do however have a growing number of innovative designers, craftspeople and woodworkers who are creating their own wood culture, mainly using hardwoods. The advantage of hardwoods – unlike softwoods – is a relatively small supply goes a long way as Irish designers and makers are proving. However, they need greater long-term continuity of supply which can only be achieved by a sustained mixed species afforestation programme, matched by support for small-scale sawmills that specialise in hardwood processing.

Meanwhile, the objective is to achieve a balanced wood industry combining innovation and carbon-friendly solutions for sustainable living regardless of scale and species; from the specialist craft and design sectors to the high-volume-demanding construction, energy and industrial markets.

Construction

While engineered or mass wood is set to transform large-scale building projects, conventional solid wood is still a major component in small-scale housing both structurally and in furniture, flooring, panelling and other fittings. This is especially true of Ireland's forest products sector. Sawn wood in conventional housing is a major Irish and UK market for home-grown

Auditorium Wexford Opera House: walnut satisfied acoustic and aesthetic criteria set by the OPW assembled design team including architects, theatre designers and acoustic consultants.

softwoods, but the potential to increase market share in timber frame construction in housing and low-rise apartments is enormous in Ireland. Log forecasts from Irish forests demonstrate there is sufficient supply to fulfil future timber frame demand as Ireland is only beginning to explore this market, compared to Europe.

Timber frame construction differs from "all timber" buildings such as log cabins as they combine wood with other materials but timber is the dominant element, significantly reducing dependence on fossil-based materials.

Timber frame buildings use wood as the main structure as well as other elements (Figure 6.3) but the finished structures are usually faced with brick, stone or rendered blockwork so they look like conventional buildings. As timber is prefabricated off site, timber frame construction is a fast, safe, clean and efficient building system.

One of the biggest advantages of timber frame is its potential in decarbonising the construction sector, which is a major

GHG emitter. This is particularly relevant to Ireland because of the enormity of the challenge to build sufficient houses by the middle of this century. A Forest Industries Ireland (FII) report compiled by Des O'Toole (2021) estimates that 581,000 housing units (single houses, scheme houses and apartments), will be required in Ireland (2022-2040) which is forecast to increase to 881,000 units by 2050. This is broadly similar to the *Project Ireland 2040: National Planning Framework* (2018), which targeted "the delivery of 550,000 additional households to 2040".

O'Toole outlines the climate change benefits of increasing timber frame construction especially in housing and low-rise apartments as well as exploring the use of CLT in medium-rise

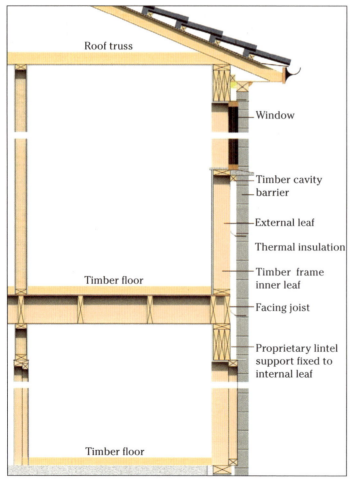

Roof truss

Window

Timber cavity barrier

External leaf

Thermal insulation

Timber frame inner leaf

Timber floor

Facing joist

Proprietary lintel support fixed to internal leaf

Timber floor

Figure 6.3: *Timber frame houses and apartments use wood as the main structure but can be faced with brick so they look like conventional buildings.*

apartments. As higher apartments are permitted in wood and CLT is introduced to the market, carbon savings become even more pronounced. However, unlike other EU member states, building taller than three storeys is prohibited in Ireland but O'Toole believes this has to change if Ireland is to maximise its timber resource and carbon savings:

Ireland plans to increase timber frame house construction from 24% to 80% of the housing market, mainly using home-grown timber.

Potential carbon savings of 3.4m tonnes CO_2 eq are achieved over the period to mid-century when comparing the base case against a market where timber frame penetration achieves an average of 55% over the period and the regulatory environment permits the use of CLT for buildings over 10m, accounting for 70% of all new apartment developments in the future.

He estimates the CO_2 savings as being equivalent to a "5.6% reduction in Ireland's current annual emissions of 60 million tonnes CO_2 eq". Only 24% of all Irish houses were timber frame in 2020 but plans to increase this market to 80% have been outlined by Coillte and FII. This target is realistic as Scotland, which has a softwood forest culture like Ireland, builds 75% of all homes in timber frame.

For Ireland, timber frame construction provides major market opportunities for home-grown sawn wood processed by sawmills and wood-based panels manufactured by board mills. These mills are strategically located in Ireland while opportunities for synergies in research and development exist between industry and other stakeholders, especially third-level colleges which have faculties in forestry, architecture, engineering, product development and design.

Attitudes towards timber frame building are changing, thanks to promotional work by FII and the Irish Timber Frame Manufacturers Association (ITFMA) which, encourages architects, engineers and specifiers to choose the timber frame method of construction. ITFMA is the representative organisation for the timber frame industry throughout Ireland with a membership of 16 manufacturers and a number of supply chain members. ITFMA members have built a number of major housing projects both in Ireland and the UK. For example, Cygnum Building Offsite, the Co. Cork- based company, built the Goldsmith Street social housing development scheme for Norwich City Council. This comprised 105 ultra-low-energy homes where the use of timber frame was a critical element in achieving the internationally recognised energy performance Passivhaus

standard. Tenants enjoy the benefits of annual fuel bills of as little as £150 (€165), which helps Norwich City Council's aspiration of tackling the issue of fuel poverty. Designed by architect, Mikhail Riches, it was the first social housing scheme to win the coveted Stirling Prize in 2019 and the Wood Awards Ireland International Award in 2020.

While Ireland has a distance to travel before creating a wood culture comparable with other European countries, Irish timber processors have made major progress not only in Ireland but also in capturing greater market share, especially in Britain. These international-scale mills have the capacity to process the increased volumes forecast from Irish forests until 2050. Irish sawmill and wood-based panel plant exports continuously demonstrate their ability to compete successfully against their European counterparts.

Goldsmith Street, a social housing development of 105 ultra-low-energy homes for Norwich City Council, is a highly innovative and sustainable timber frame housing scheme. It was the first social housing scheme to win the coveted Stirling Prize in 2019 and the Wood Awards Ireland international category in 2020. Built by Cygnum Building Offsite, the Co. Cork-based company, timber frame was a critical element in achieving the Passivhaus standard.

Sawmilling

The modernisation of Irish sawmilling began in the late 1980s and has continued apace since then. In 2023, five sawmills processed approximately 95% of timber production in Ireland with capacity to increase this in line with forest yield and production which will double to eight million m³ by 2035 (Figure 6.4).

All Irish sawmills are family-owned businesses, deeply rooted in their communities. By the early 2000s, sawmills had rationalised and invested heavily in new technology such as

Security of timber supply to Irish sawmills and panel board mills requires a sustainable forestry programme supported by a viable harvesting (below) contracting and haulage network (right).

machine grading, kiln drying and highly automated cutting lines. Ciaran O'Connor, State Architect, sums up the development of Irish sawmilling since the 1980s:

> The story of Ireland's sawmills is impressive. When I returned from Canada in the 1980s and worked on the innovative first phase of Killykeen Forest Park – which featured the experimental use of six Irish grown timber species – the country had some 120 sawmills. None of these had combined drying and mechanical stress grading facilities. Today Irish sawmills have been rationalised and automated to compete successfully with their European counterparts. Now, we have unified European timber standards for timber quality, structural grading and preservation.

Sawmills also carried out strong marketing campaigns in the UK, supported by Enterprise Ireland, while Coillte, the main source of logs until recently, also played a part in guaranteeing a continuous supply of timber regardless of market and price fluctuations. Despite an outdated State felling licensing system, which has slowed down log supply especially between 2020 and 2022, Irish sawmills have a sustained indigenous timber source and so can guarantee a sustained supply of products to Irish and UK traders.

Irish sawmills have the capacity to increase production in line with forest yield which will double to eight million m³ by 2035.

Sawmills increased exports to the UK from 20% to 70% after the collapse of the domestic housing market. This robust export strategy was vital to their survival as Irish construction went into freefall from an average of 79,635 dwellings built in the five years preceding the economic crash to an average of 5,843 during the five-year period 2011 to 2015 according to Central Statistics Office (2022) and Department of the Environment, Heritage and Local Government (2023) data. Yet, not one sawmill or board mill went out of business after the most devastating recession in the history of the State, demonstrating the resilience and sustainability of timber processing in Ireland.

In addition, sawmills pay high prices for sawlog material to both Coillte and the private sector – mainly farmers. In many instances prices paid for 30-year-old Sitka spruce and other conifers are higher than prices achieved by European forest owners for pine and spruce grown over 80-year rotations.

Although sawmills are heavily reliant on construction, they also produce pallet wood for the packaging market, fencing and specialist wood products for the leisure and garden sectors. An estimated 54% of the log ends up in sawn products and the remaining wood residue – wood chips, sawdust and bark – is further processed into a wide range of end products including wood based panels, energy and horticulture. Residues are used

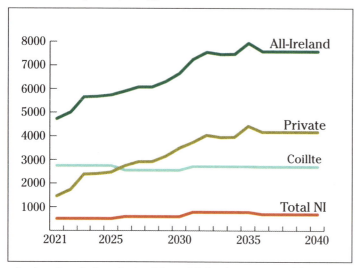

Figure 6.4: *Forecast of potential volume production by forest ownership type 2021-2040 (000 m³ overbark).*

Source: Adapted by D. Magner from All Ireland Roundwood Production Forecast 2021-40, (2021). COFORD, Department of Agriculture, Food and the Marine, Dublin.

Note: It is estimated that 94% of all production in Northern Ireland is sourced in Forest Service, Department of Agriculture, Environment and Rural Affairs forests.

for heating timber-drying kilns with further potential for energy generation in combined heat and power plants (CHPs) which sell energy to the national power grid. Some sawmills sell wood chips and wood pellets to businesses who wish to convert from oil and gas to renewable energy.

Wood waste to renewable energy

The wood energy market has yet to be fully explored in Ireland unlike a number of European countries where half the renewable energy market depends on bioenergy or biomass from agricultural, waste but mainly sourced from wood.

What is it about Ireland and wood energy? Not one political party has championed this renewable energy resource. Yet, it leads the way in global renewable energy consumption (Figure 6.5). Countries, such as Denmark, Sweden and Austria, are well underway in achieving renewable energy targets. They regard wood biomass as an essential part of their bioeconomy and climate mitigation policies.

Although Austria enjoys huge advantages in renewable energy – especially hydropower – over other EU countries, it places a high value on wood energy. Hans Kordik summed up the Austrian position: "While Austria's renewable energy mix is made up of hydro power, solar thermal energy, photovoltaic,

Approximately 95% of all timber production on the island of Ireland is processed by the following sawmills: ECC, Glennons (pictured), GP Wood, Laois, Murrays, CJ Sheeran and Woodfab.

geothermal and wind energy, more than 50% of the [country's] renewable energy mix comes from biomass."

When discussing wood biomass, it is easy to fall for the myth that burning wood has a negative impact on greenhouse gas (GHG) emissions. Kordik, a senior agricultural consultant with the World Bank dismisses this because "as long as the forest acreage stays the same or – as in the case in Austria – even grows, biomass is carbon-neutral". Denmark, where the renewable energy is divided almost equally between wind and biomass, adopts the same position. In Sweden, wood biomass provides approximately 60% of the fuel for district heating and municipal waste provides another 15%.

Yet, Ireland has, at best, an ambivalent approach to wood energy even though our forests produce far higher yields than Scandinavian forests. So, while the main use of timber is for construction, furniture, fencing, panel board, packaging, etc., wood energy should be an inherent part of the renewable product mix. Despite recent poor afforestation rates, the Irish forest area has increased at a far higher rate than other EU countries. Although there are significant cost savings and major climate change benefits, the reaction to Ireland's Support Scheme for Renewable Heat (SSRH) has been cautious. Initially, conversion to wood from oil and gas was extremely slow but this is changing as more businesses are switching to wood energy. While the general consensus is that scheme approval is overly onerous and long, the SSRH is acknowledged as an excellent initiative. The savings are considerable for investors who wish to transfer from fossil-based fuels to wood biomass based on best available information in 2021 (Table 6.1).

When the SSRH was launched in 2019, the energy sector believed that Ireland would begin to catch up with Austria. Denmark, Sweden and other EU countries. "The introduction of the SSRH is a game changer as Ireland now urgently needs to address its declared climate emergency," said the Irish Bio-energy Association (IrBEA) then president, Des O'Toole. The uptake of the SSRH scheme has been disappointing, despite major savings for businesses that convert to wood.

There are a number of rural communities that would benefit enormously from wood energy. For example, Co. Leitrim-based companies McMorrow Haulage & Firewood – near Dowra – and McCauley Wood Fuels, Mohill, have the capacity and expertise to convert waste wood into renewable energy. Like other companies they are members of the IrBEA's Wood Fuel Quality Assurance (WFQA) scheme which ensures that wood fuel is sourced from sustainably managed forests and chipped to consistently low moisture levels.

McMorrow has the capacity to supply 100,000m³ of wood chip annually in the medium term with the potential to double output if the SSRH scheme gathers momentum. McCauley is a smaller supplier at 12,000m³ annually but could increase output to 50,000m³ of wood chips in the short term to businesses in Leitrim and neighbouring counties. Customers include leisure centres, large-scale mushroom and pig farms, nursing homes and hotels.

At a capacity of 150,000m³ (1.5 % of annual Irish forest production by mid-century) both McMorrow and McCauley could provide continuity of wood chip supply to over 400 businesses each with an average annual heat requirement of 1,100MWh. This would displace 56 million litres of oil per annum as well as creating further employment in harvesting, processing and transporting the raw material as well as the installation and servicing of renewable energy boilers. Switching to wood energy also reduces our dependence on fossil-based energy which is why the Austrians and the Danes made the switch to renewables.

Wood waste to wood-based panel (WBP) products

While increased volumes of small logs and residue will result in increased biomass for heating, timber processors prefer to

---

RENEWABLE WOOD ENERGY – FOLLOWING THE DANISH EXAMPLE

It is often assumed that Denmark's main renewable energy source is wind. Wind generates 20% of its renewable heat and energy compared with 55% biomass (48% wood and 7% straw), 9% biodegradable waste, 4% liquid biofuels, 5% biogas and 7% other sources according to Danish Energy Statistics (2018).

Ireland has an excellent Support Scheme for Renewable Heat (SSRH), which could help Ireland to be a leader in renewable wood energy.

Based on 2024 supports, the scheme pays out on a reduced tiered rate per kilowatt hour (kWh) of heat produced, beginning at 5.66c for the first 300,000kWh; 3.02c for next 700,000kWh; and 0.50c for the next 1.1m kWh and so on until the scheme duration is complete at 15 years.

This government scheme operated by SEAI aims to "bridge the gap between the installation and operating costs of renewable heating systems" and to "incentivise the development and supply of renewable heat". The scheme provides an installation grant for heat pumps and "on-going operational support" for commercial, industrial, agricultural, district heating, public sector and other non-domestic heat users.

The savings in converting from fossil fuels to wood are considerable. For example, annual savings are estimated at €52,347 by converting from kerosene to wood chips (Table 6.1) or €44,013 by converting to wood pellets, using a heat pump with an annual requirement of 1,100MWh. For the same unit, annual savings of €31,963 can be achieved by converting from natural gas to wood chips or €23,629 if wood pellets are used.

Table 6.1: *Indicative example of costs and savings (€) by investors availing of the SSRH scheme using a biomass heating unit with an annual heat requirement of 1,100MWh.*

Fuel type	Wood chip	Wood pellet	Natural gas*	Kerosene*
Annual Heat Requirement (MWh)	11100	1100	1100	1100
Fuel Quantity and units	354 tonnes	250 tonnes	1100 MWh	107,691 litres
Cost of fuel/unit (€)	160/tonne	260/tonne	50/MWh	0.70/litre
Total fuel cost (€) X	56,657	64,991	55,000	75,384
Operating, maintenance costs (€) Y	5,000	5,000		
SSRH support for 15 years (€) Z	38,620	38,620		
Annual heating cost (€) X+Y-Z	23,037	31,371	55,000	75,384
Existing cost of boiler system (€)	200,000	180,000		
Annual saving over natural gas (€)	31,963	23,629		
Years to pay back capital – gas	6.3	7.6		
Annual saving over kerosene (€)	52,347	44,013		
Years to pay capital – kerosene	3.8	4.1		

Note: This table has been developed for indicative purposes only by the IrBEA. and adapted by D. Magner 2021.

*Fossil fuel prices can fluctuate greatly, which can impact on the economics of all energy projects. Carbon tax in Ireland in 2021 was €25/tonne. At that time carbon was forecast to increase to €100/tonne in 2030 for all fossil based fuel.

Figure 6.5: Breakdown of renewable energy systems in global energy consumption.

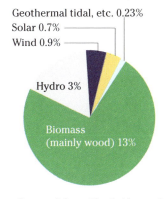

Geothermal tidal, etc. 0.23%
Solar 0.7%
Wind 0.9%
Hydro 3%
Biomass (mainly wood) 13%

Source: Adapted by D. Magner from World Bioenergy Association, 2017. Note: Renewables hold 17.8% of global energy consumption compared with oil (39%), gas (21%), coal (20%) and nuclear (2%).

make things out of waste wood rather than burn it. Small and low-value logs (pulpwood), along with wood chips and sawdust are converted into a range of products from paper to high added-value wood based panel products. This form of wood engineering existed in Europe and North America since the late nineteenth century and had developed into a major global industry by the first half of the twentieth century. Ireland's transition to this market was late – and painful – beginning in the early 1960s. It came to an abrupt end during a brief period in the late 1970s and early 1980s. A number of paper and chipboard factories closed down including Munster Chipboard, Waterford (1979), Clondalkin Paper Mills (1982) and Chipboard Products, Scariff (1983). This placed Irish forestry, especially sawmills, in an invidious position as the viability of the sector depends on having markets for wood residue and small logs.

The industry was faced with dumping wood residue and leaving small timber unthinnned – and unharvested – which would have threatened the sustainability of forests and sawmills. The ability to turn this market around, after the total collapse of paper and board mills, has been a major success story in Irish forestry. Within two decades, three international-scale WBP mills were established in Clonmel, Waterford Harbour and

Carrick-on-Shannon. Two of these are now in State ownership since Coillte purchased the OSB mill in Waterford Harbour in 2002 and the Clonmel MDF plant in 2006. These produce the Medite Smartply range of board products. Masonite Ireland, owned by Masonite International, continues to produce moulded door facings mainly for export. It is no coincidence that Irish sawmills expanded in tandem with board mills as there is a high degree of inter-dependency between all timber processors.

WBPs manufactured from low-value small logs and waste wood, are essential products in ensuring there is no waste in wood processing. Depending on the end product, timber strips, wood chips or sawdust are reprocessed by subjecting them to chemical, heat and pressure treatment. The required fibre from suitable fast-growing softwoods – mainly spruce and pine – are ideal, and readily available, for producing MDF, OSB and hardboard.

Masonite International manufacture a wide range of products based on the original Masonite hardboard, patented in 1924. Established in 1996, Masonite Ireland concentrates on the production of a durable hardboard for moulded door facings, mainly for export. Strategically it was vital for Coillte, private forest owners and sawmills in the west of Ireland to find markets for small logs and timber residue. In addition, it is a major employer in Co. Leitrim where it provides 151 direct jobs in the region with additional employment in services and industries that have been created since its establishment such as paint manufacturing and renewable energy generation.

Medite Smartply, owned by Coillte, and based in Clonmel, produces MDF and OSB mainly for the export market. Compared with chipboard or particleboard, MDF has a more homogeneous finish and has excellent machinability. As a result it can be veneered, painted, glued and doweled. It can also be combined with other material including solid wood, steel and glass for use in furniture and panelling.

Medite's MDF plant has expanded its product range in recent years from general-purpose products to boards that resist moisture, flame and harsh weather conditions. It produces durable panels that provide dimensional stability "which allows it to be used in applications once limited to products such as concrete, plastics or metals" the company states. This allows it to displace fossil-based materials while other Medite products, which combine acetylated wood fibre with manufacturing technology take engineered wood to a level where it can displace scarce tropical hardwoods. Medite Smartply produces OSB in its Waterford Harbour plant. OSB has a major role to play in timber frame and low-energy construction. OSB is suitable for load-bearing applications and has increased its market share in the structural panel sector, even against plywood. It is used in wall sheathing and flooring in timber frame construction. OSB has expanded its product range and finishes so it is now suitable for use in humid conditions.

ACCOYA AND TRICOYA DISPLACING TROPICAL WOOD

Building tall in engineered wood has huge potential to displace fossil-based materials but technological treatment of softwood from sustainably managed forests can also be used to displace scarce hardwood timber resources from threatened forests. For example, heat treatment of timber to produce products such as Accoya and Tricoya now provide softwoods with the longevity and decay resistance of tropical hardwood species such as mahogany.

Accoya wood is highly rot-resistant and stable regardless of climate. It is chemical-free in the manufacture process of wood acetylation, while it requires no preservatives, even when used externally for doors, window frames, decking and outdoor cladding.

Tricoya such as Medite Tricoya Extreme has exceptional durability, moisture resistance and dimensional stability (pictured). It is supported by a 50-year guarantee. Like Accoya, it can be used externally and withstands extreme cold and wet weather conditions.

MDF, manufactured from Sitka spruce predominantly, is processed at the Coillte-owned facility in Clonmel. It has a wide range of panel uses such as here at the European Bank for Reconstruction and Development.

Overleaf: Designed by architects Clancy Moore, Taka & Steve Larkin, the Big Red installation, at the London Festival of Architecture 2015, featured OSB and sawn Irish timber.

Engineered wood in Ireland

Ireland has been extremely innovative in developing an indigenous WBP industry, but it has yet to venture into engineered mass timber construction. This provides the greatest opportunity to build apartments and multiuse buildings as well as displacing fossil based materials. To enter this market, architects and builders need a sufficient supply of raw material, which Ireland has, and planning laws that allow them to design medium- to high-rise buildings. Ireland's planning laws prohibit building above three storeys in mass timber, unlike all other EU countries but this is finally being addressed. Ireland has a sufficient supply of spruce but has it the strength and bending characteristics required for large-scale projects? Irish timber is mechanically graded to an EU standard C16, which is a widely used timber grade. Research work carried out in University of Galway shows that home-grown Sitka spruce is suitable for CLT production. Dr Annette Harte outlined her findings in 2017:

> As the technical viability of producing CLT panels from Irish grown Sitka spruce has been established from a structural point of view, the next natural step in moving towards commercialisation is the identification of a producer that will bring this to large-scale production.

The potential to maximise this market in the long term adds to the spruce market range and presents major economic and environmental benefits in sustainable construction in Ireland. It will, however, present a challenge to the Irish timber processing sector to manufacture CLT but first "It will require a change in Ireland's building regulations to allow buildings higher than 10m to be built primarily in CLT", as outlined by FII. While Ireland is unlikely to build high-rise apartments or offices anytime soon, mass timber use in low, to medium-rise structures can play a major role in achieving net zero by 2050. Table 6.2 demonstrates the difference in net carbon emissions between a hypothetical conventional tower block and a CLT equivalent, comprising 48 dwellings (24 one-bedroom apartments and 24 two-bedroom apartments) or office/retail space equivalent, over six storeys. The increased timber usage is also displayed.

Buildings of this scale can make a major contribution to decarbonising the construction industry as Richard Romunde of Gresham House points out, quoting from 2019 EC data (Wood-Building the Bioeconomy):

> The carbon saving figures of using wood can be substantial. The city of Helsinki built four similar five-storey apartment blocks, two in wood, two with concrete. The production of materials used in the timber buildings had a 74% lower carbon footprint

Ireland has yet to even acknowledge the importance of building in mass wood, unlike France which has enacted a law requiring at least 20% wood or sustainable products in public buildings from 2022. This is achievable in France, which has 31% forest cover. Even the Netherlands, which has a lower forest cover than Ireland, has ambitious plans for renewable construction as Amsterdam has mandated that 20% of all new housing projects must be constructed with wood or other bio-based materials from 2025.

When asked about the potential for future expansion of CLT manufacturing for countries like Ireland and the Netherlands, Dr. van der Lugt told the *Irish Farmers Journal* that this would require a sustainable supply of timber which is not possible in the Netherlands because of its low productive forest cover. "Investment in CLT manufacture is high especially in grading lines and gluing technology while a sustainable supply of at least 200,000m^3 would be required with an output of over 100,000m^3 of sawn material per year." Four Irish sawmills process over

Table 6.2: Carbon balance comparison of a six-storey tower block.

Construction material	Net emissions (tonnes CO$_2$)	Timber m^3
Concrete	1,016	202
CLT	-437	1,205

Source: The BioComposites Centre, Bangor University and GHAM forecasts, 2020.

Sample of Sitka spruce CLT, University College Galway. Post doctoral researchers at UCG have developed and tested a viable CLT product made from Irish-grown Sitka spruce, which meets the structural requirements of current design standards for large-scale construction.

300,000m³ annually with the capacity to increase this as supply will double by 2040. The results of research carried out by University of Galway demonstrates that Sitka spruce is suitable for CLT production. With timber production on the island of Ireland forecast to increase to 7.5 million m³ – mainly spruce – within 10 years, the establishment of a CLT plant may be the next strategic move by the Irish timber processing sector. Investment will prove challenging, but timber supply will not be an issue.

While Ireland's antiquated planning laws militate against mass wood construction, innovative architects and engineers are using CLT and glulam in a range of buildings.

CLT was the main engineered wood used in the winning entrants for the 2014 and 2016 Wood Awards Ireland (WAI) by Bucholz McEvoy Architects. The Samuel Beckett Civic Campus and Ballyogan Maintenance Centre are both located in south County Dublin. What is also encouraging about these two public buildings is the role played by Dún Laoghaire-Rathdown County Council (dlr) in encouraging collaboration between Bucholz McEvoy Architects and the Council's dlr Architects.

Center Parcs Ireland, located near Ballymahon, Co. Longford, demonstrates the aesthetic structural qualities of mass wood in large-scale building projects. Completed by John Sisk & Son (Holdings) Ltd. with Holder Mathias Architects and Peter Brett Associates, this leisure project explores wood in all its forms, from solid timber to engineered softwood including glulam and CLT. Overall winner of Wood Awards Ireland in 2020, the resort includes 466 self-catering lodges and 30 apartments. All the wood used was Forest Stewardship Council (FSC) and Programme for the Endorsement of Forest Certification (PEFC) certified. Imported wood was in rough sawn format which was then manufactured in an Irish sawmill to achieve several different sizes. Species used, included Norway spruce, European and Siberian larch and western red cedar with Elliotis pine plywood, SafeStrip AntiSlip Decking in softwood (redwood) and hardwood (Bangkirai).

While these large-scale office and leisure buildings demonstrate the importance of wood as a building material, so far most of the wood used in these Irish projects was sourced from sustainably managed forests overseas. This is slowly beginning to change as architects see the merit of using homegrown timber for large buildings. In the 1990s, Coillte decided to build a new regional office in a forest close to the village of Newtownmountkennedy, Co. Wicklow, using over 90% homegrown timber. Designed by the architect Duncan Stewart, it was the first all-timber office complex in Ireland when it was completed in 1997. It was such a success, Coillte decided to enlarge the building and move its corporate headquarters to Newtownmountkennedy. The building is heated by a wood-fuelled boiler from wood residue sourced in Coillte's forests. Species used for construction include Sitka spruce, Norway spruce, European larch and Douglas fir while oak, ash and elm were used for furniture and internal fittings.

Dún Laoghaire Rathdown County Council Operations and Maintenance Centre, Ballyogan, Co. Dublin, is a low-energy naturally ventilated building constructed in engineered glulam wood. The design has optimised passive solar gain, controlled to prevent overheating by the use of automatically and manually controlled blinds, windows and timber vertical ventilator panels. Approximately 1,000 tonnes of CO_2 are sequestered within the timber structure.

CREATING IN WOOD

Different species such as oak and spruce can have their own design language that speaks of their own physical attributes. From ancient shelters in forest clearings, to traditional craft and innovative design, wood has shown itself to be a natural, authentic and ecological material capable of great pragmatic and poetic beauty.
– Ciaran O'Connor, State Architect, *Wood Awards Ireland 2014*

As we have seen, the emphasis in Ireland has been to create a major softwood resource. This is understandable, as wood security is a laudable objective and softwood is the only cost-effective and practical solution for construction. As more better-quality land becomes available for forestry, a greater range of options will be available to create a balanced timber resource which accommodates conifers and broadleaves

Center Parcs Ireland, located near Ballymahon, Co. Longford demonstrates the structural and aesthetic qualities of engineered wood in large-scale building projects.

Overleaf: Boardroom, Coillte HQ, Newtownmountkennedy, Co. Wicklow. Home-grown timber was used throughout the building. The chairs and carved boardroom table are made from sycamore and elm, which blend with the arched spruce ceiling supported by glulam beams.

– softwoods and hardwoods – each as Ciaran O'Connor says with its "own design language".

Irish furniture and craft designers and makers opt for hardwoods and softwoods from specialist coniferous species. Ireland's forests and woodland produce a minimal supply of hardwoods because the increased plantings of mainly native species took place only since the early 2000s. These plantations won't reach maturity until the second half of this century, while the large ash afforestation programmes will need to be replaced due to ash dieback.

According to *Forests Statistics Ireland 2022,* only 26,000m^3 of hardwoods were harvested in 2021. Some 60% of this resource was used for firewood and pulpwood so less than 9,000m^3 was suitable for furniture, craft and turnery. However, small-scale mills and workshops operate successfully with a limited supply of homegrown hardwoods and many also use some softwoods which have potential for furniture, panelling, craft and other high added value end uses. These are often referred to as "minor conifers" and include Scots pine, larch, Douglas fir, western hemlock, western red cedar and yew.

While these are scarce, Ireland has 44,400ha of specialist softwoods and 66,100ha of potentially suitable hardwoods, after discounting small broadleaves (willow, mountain ash, etc.), ash and most birch planted prior to the advent of "improved" nursery stock. It is likely that timber traders and woodworkers will continue to supplement supply by importing from countries that comply with the EU Timber Regulation (No. 995/2010).

Sawmills, suppliers, designers and woodworkers, who specialise in hardwoods and specialist softwoods, need support to develop this resource in areas such as kiln drying, finishing and presentation so they can compete successfully with imports. This approach will allow a new wood aesthetic to emerge which is seen to best effect in construction, furniture and interior design as well as restoration projects where traditional wood working is refashioned to meet contemporary needs.

Furniture and interior design

Nowhere is the use of wood so deeply rooted in our psyche as in the home, especially in furniture. It is impossible to think of any other material that is so physically and psychologically inviting, with the timeless quality to transform the mood of a room or a building, and the mood of the occupants. It is the touchstone to begin and close each day as window frame, bed frame, picture frame, cupboard, door, table, chair or kitchen utensil. Wood is our daily reminder of its place in our collective culture.

Its strength, durability, beauty, redolence and versatility mark it apart from other materials. Its renewability means that after harvesting, it is replaced, to ensure that the growing and production cycle remains intact, so that wood continues to live in our households and in the forest. Its versatility allows woodworkers and designers to shape and profile it so that it is both structurally strong and lightweight. The most positive aspect of current design is the number of Irish designers that are embracing wood as an innovative contemporary medium.

The future of wood design in Ireland is healthy, thanks to a number of developments, not least support and promotion by the Design & Crafts Council of Ireland (DCCI) and the establishment of craft and design bodies at county level. DCCI promotes the commercial development of Irish designers and makers, many of whom work in wood. They are producing furniture, wood-turned objects and artefacts that explore new horizons in wood design.

Many of these are taking wood beyond its traditional boundaries where the lines between art and craft, utility and virtuosity are continuously being redrawn. Joseph Walsh, founder of Joseph Walsh Studios in Riverstick, Co. Cork takes wood in a new direction by engineering it into curvilinear shapes through lamination. The result is a triumph of both imagination and skill. Ash is the species of choice for Walsh, although he works with other woods and material, often combining them.

Magnus Modus, *National Gallery of Ireland, by Joseph Walsh Studios.*

Magnus Celestii, *olive ash Joseph Walsh Studio. See also pages 208 - 209.*

Ash has strength and elasticity which allow Walsh to push the boundaries of design to produce furniture that unites function and aesthetics. For example, his piece *Magnus Celestii* lives up to its "great heavenly" claim as this free-form, made from layers of laminated ash, is both furniture and sculpture. If there is one aesthetic or artistic answer to "Why Wood?" *Magnus Celestii* provides it.

When *Magnus Celestii* won the Wood Awards Ireland Innovation Award in 2016, Ciaran O'Connor, chair of the judging panel described it as "an extraordinary installation combining vision with ingenuity".

Walsh and his team of master craftsmen transform spaces as diverse as churches, corporate boardrooms, artists' studios, bedrooms and art galleries. While he works in a wide range of mediums, his signature pieces are in ash or olive ash which is the heartwood from ash and resembles olive wood in appearance. It is precision-sawn by a skilled French sawmiller and then veneered. When I interviewed him for an *Irish Farmers Journal* feature in 2016 he said the sawmill operators "examine trees with an eye to craftsmanship rather than looking at volume running through the mill [as] they are sensitive to the wood's characteristics and saw it in the best possible way".

During that interview I received a glimpse of his creative thought process when designing works. The whole process begins with some free sketches, followed by model making. He showed me a model of a church in Verona and his design for a large-scale wall piece – *Magnus del Busolo* – spanning over 20m, which has since been installed and described as:

> … a permanent artwork conceived for the sacred space of the 14th-century church of the family of Conti da Lisca di Formighedo. It was designed in response to the patrons' desire to commission a new work of contemporary artistic expression, both beautiful and functional, for the church which their family has maintained and enhanced for centuries.

Magnus del Busolo, *laminated ash in the church of the family of Conti da Lisca di Formighedo, Verona, Italy.*

Walsh works with European ash, which is the same species as native Irish ash. He would prefer to use home-grown ash, but relies on imports from France simply because the required material is not available in Ireland.

The tragedy for Ireland is that the increased ash planted between 2000 and 2012 – due for harvesting as early as 2030 – will never be sawn by Irish sawmillers and shaped by Walsh's craftsmen as it has been wiped out by ash dieback; its revival depending on research into developing a disease resistant strain of ash.

In addition to Joseph Walsh, a number of Irish designers and architects are using wood to redefine its living environment. Wood is the ideal material to transform space – and place – because it has the connectedness that links the dweller with the organic: the medium is the message.

Furniture designer Alan Meredith bends and shapes wood into a diverse range of refined intimate compositions including

Joseph Walsh and his team of master craftsmen transform living, sacred and secular spaces in works such as Enignum VI *(above) and* Enignum VIII, *Devonshire Collection, Chatsworth House.*

chairs, tables, shelving and wood-turned objects which are both functional and sculptural objects. He utilises the steam-bending technique by reshaping home-grown hardwoods, whether they are yielding like ash or rigid such as oak. The wood is shaped into refined, sophisticated and original compositions such as his "Vinculum Series". This approach allows him to produce a unified design, whereby:

Vinculum console drawer, part of Vinculum *series* by *Alan Meredith*

Contemplation Space *in the Mater Hospital, designed by Garvan de Bruir, Laura Magahy and Michael Goan with John Sisk Training and Education Centre providing the structural work.*

Table surfaces become table legs while a single plank of oak travels through 180 degrees to become two levels of a shelf. Through the process of reading the properties of the material and exploring its possibilities, new and original forms emerge.

Meredith's approach permits him to dispense with artificial treatment. Instead, he opts for scorching, fuming and white oil finishes.

Contemplation Space – winner of the Wood Awards Ireland innovation category in 2014 – located in the foyer in the Mater Hospital, Dublin, uses ash to provide a unique resting and meditation space in the busy foyer. Even though the passerby can partially see in and the occupants can see out, privacy is preserved by the introduction of an additional layer of ash lattice work at eye level to provide increased visual privacy and structural stability, while maintaining the integrity and beauty of the overall form which can be enjoyed by visitors and hospital staff.

Overleaf: Carmelite Prayer Room, Clarendon Street Priory, off Grafton Street, Dublin by Niall McLaughlin Architects.

Ash is also the species used in The Carmelite Prayer Room in Clarendon Street Priory. This time spirituality is combined with intimacy. The design is also conscious of the acoustic role of wood to ensure the contemplative integrity of what is a place of prayer and meditation. A special award winner in the 2014 Wood Awards Ireland, ash is used innovatively to remodel and restore an earlier prayer space that Ciaran O'Connor said "is both powerful and ephemeral ... in terms of its identity and wonder".

Irish wood turners have a strong tradition of creating functional objects but in recent years they have pushed the boundaries of wood turnery to a new creative level, none more so than Emmet Kane. His 2015 solo exhibition, in the National Museum of Ireland, Decorative Arts & History, was a tour de force in creative woodworking. The exhibition explored Kane's remarkable journey of development as an artist and wood turner, featuring a diverse array of work including functional vessels and bowls, wall hangings, artistic pieces and small-scale intimate works. He works in more than a dozen species including native oak, cherry, elm, holly and apple which he textures, ebonises, gilds and colours to produce vibrant pieces, but he never loses sight of the wood's unique characteristics.

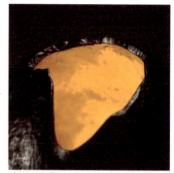

Above: Gold Tulip, ebonised oak with gold leaf, Emmet Kane.

Alan Meredith, as we have seen, specialises in bespoke furniture, but also transitions seamlessly to sculptural wood-turned vessels. Meredith's works are turned and hollowed from solid unseasoned oak. He says:

> The intention is to create wholesome and strong forms that reveal the qualities and strengths of the oak, which has a robust character and is malleable when steamed. Through the sympathetic manipulation of the material one hopes to explore new possibilities and original forms. The solid oak pieces are turned on the lathe to a thickness of 3mm. This transforms the oak into a flexible and pliable material which, once steamed, can be manipulated into new and exciting forms.

Oak Vessels, Alan Meredith.

Once the oak dries it becomes hard again and holds its new form. Great skill is needed to ensure an even-wall thickness, which is essential in creating a balanced and lightweight work.

RESTORATION AND CONSERVATION

While wood survives for centuries, it eventually decays but contemporary designers and woodworkers can restore – or

Oak shelving from the "Vinculum Series" by Alan Meredith, Alan Meredith Studio.

recreate – it to live on. When Kilbeggan Distillery in Co. Westmeath, founded in 1757, wished to reconnect with its original distilling method, it needed to restore its waterwheel which had lain derelict for decades. This was achieved through collaboration between the distillery, a sawmill and a foundry. Industrial archaeologist, Dr Fred Hammond, Athy Foundry and Sheehan's Sawmills, Ballyporeen in Co. Tipperary combined in the restoration which included research design and sourcing home-grown European larch to rebuild the wheel floats while oak was used in the sluice gate.

The restored waterwheel in European larch and oak sluice gate, Kilbeggan Distillery, Co. Westmeath.

Restoration and reuse were dual objectives by McCullough Mulvin Architects for its restoration project in the Dublin Dental Hospital. The architects combined American white oak with existing pine floorboards to create a unified space without damaging the integrity of the existing protected buildings in Trinity College. The concept was, according to McCullough Mulvin "about inhabiting forgotten rooms of varying scale in the city, with a view to transforming the potentiality of a series of ordinary city structures into a richly layered internal environment". The project won the Wood Awards Ireland conservation award in 2014.

Restoration of the Dublin Dental Hospital by McCullough Mulvin Architects using oak combined with existing pine floors.

WOOD VOYAGE

Wood conservation is best expressed in Ireland in boat restoration projects which demand research, skill, innovation and a sense of history. The upsurge in boat and ship building is influenced by a past history of famine, invasion and trade. This is expressed in projects such as the Dyflin Viking Ship and the Viking Longboat along with emigrant ships such as the Dunbrody and Jenny Johnston. The Meitheal Mara in Cork City and The Big Boat Build project, Skibbereen, Co. Cork, run by the AK Ilen Company is a not-for-profit charitable trust, which carries out educational workshops in boat-building.

A number of boat restoration and recreation projects have been carried out in Co. Clare which illustrate the capability of boat builders in the county. In Querrin, the Shannon Hooker, known as the *Sally O'Keeffe* – in memory of the woman who sailed her – was reimagined by naval architect Myles Stapleton while keeping the integrity of the original design intact. Led by master shipwright, Stephen Morris, the project involved local tradespeople and apprentices, with research by Críostóir Mac Cárthaigh. All the wood used was home-grown, including European larch for the planks and oak for the keel and ribs.

Again, Stephen Morris was involved in the restoration of the yacht *Naneen* along with naval architect, Paul Spooner and project manager, Fionan de Barra. *Naneen* was one of a fleet of seven classic, gaff-rigged yachts built in Ireland between 1903 and 1908.

The hull of Naneen *was constructed upside down. Laminated Douglas fir, iroko and mahogany were the main species while salvaged pitch pine was used in the cabin and cockpit floor boards.*

All the wood used in The Sally O'Keeffe *was home-grown including European larch for the planks (top left), oak for the keel and ribs (top right), before it was finished and set sail in 2012.*

Designed by the naval architect, Alfred Mylne in 1902, for members of Dublin Bay Sailing Club, the fleet raced in Dublin Bay until 1986 when major restoration was required. The fleet was laid up in Arklow until a restoration project began in 2017 with the formation of The Dublin Bay 21-Footer Classic Yacht Association whose members transported *Naneen* to Kilrush boatyard.

While the *Sally O'Keefe* is a re-creation, *Naneen* was an authentic wooden boat restoration initiative, using Alfred Mylne's original drawings supplemented by construction details by Paul Spooner. The original two-ton lead ballast keel was reused together with the original iron tiller and fittings and some greenheart and pitch pine from the original vessel.

Central to the project was the combination of the traditional skill of the shipwright and the application of the latest technical knowledge in timber conservation and innovative wooden boat construction. The result illustrated the effectiveness of wood as a structure and as a lightweight skin, capable of withstanding the toughest of marine environments. The innovative use of laminated beams and frames and epoxy resins was combined to create a stiff, water tight, low maintenance, monocoque hull, without nails or screws, which allows the use of durable polyurethane finishes. *Naneen* is the first of the fleet to be restored. At the time of writing, three more vessels are being built at Kilrush Boatyard, led by Stephen Morris, who employs local tradespeople and apprentices.

Naneen sets sail after 3,600 hours of restoration work.

Traditional and contemporary timber use combine to provide a multiplicity of answers to Why Wood? but the common themes of renewability and sustainability are ageless. We know that for every tree harvested in a sustainably managed forest, the surrounding trees compensate by increasing their growth and carbon sequestration. In the case of final harvest, each tree is replaced by at least three saplings – manually planted or self-regenerating. The chain doesn't end in the forest but is prolonged in end use and reuse which is why Ireland's Climate Action Plan has identified forests and forest products as key elements in achieving carbon neutrality by 2050.

This ambitious target can only be achieved if Ireland has sufficient forests to sequester CO_2 and enough sustainable wood to displace fossil-based materials in construction, packaging, energy and other carbon-intensive products identified by countries with strong wood cultures. Irish wood processors, designers and makers work with the tree species available: we need to work with what we have and prepare for the increased volumes coming on the market. We need to increase our hardwood resource but also to maximise the value of our flexible and versatile softwood resource. In the meantime, we need to approach species such as oak and ash as wood with "their own design language that speaks of their own physical attributes," explains Ciaran O'Connor.

This approach requires protecting and enhancing the few remaining semi-natural forests that act as a counterweight to the commercial fast-growing coniferous forests. It also requires innovation and openness to the possibilities of creating a diverse wood design paradigm "capable of great pragmatic and poetic beauty", maintains O'Connor. The restoration of the oak roof of Carrickfergus Castle, Co. Antrim and the design, and

making of *Vessel* – in Sitka spruce – for the Venice Biennale illustrate O'Connor's philosophy.

Oak was the initial preferred species by O'Donnell + Tuomey Architects for their stunning *Vessel* – a wood-stack temporary structure – as a special installation for the Venice
Architecture Biennale, 2012. The structure was meticulously
assembled from 5,250 pre-sawn panels of Sitka spruce sourced
in Coillte forests, processed by Glennon Brothers Sawmills and
manufactured by Gem Joinery. The construction of *Vessel* became an experiment in the design control of Sitka spruce – the
workhorse of Irish forestry. The end result demonstrated the
aesthetic potential of a raw material not previously recognised
for its fine detail. *Vessel* with its 9-m high sculptural light funnel
rising from a 4m x 4m floor plan was a pivotal element at the
Venice Biennale, where it attracted 178,000 visitors. Both Carrickfergus Castle and Vessel are contrasting projects in time
and space but their juxtaposition in the story of wood is justi-
fied as they are a reminder of the practical and poetic expression of wood as a sustainable building material. In both
instances, they are the minor but essential materials in spaces
dominated by brick and stone.

The restoration of Carrickfergus Castle, as already discussed, shows what can be achieved using a scarce resource
in a creative and sensitive restoration. The project team located windblown oaks which feature in the open oak truss design. The ridges of both hipped roofs were topped with
louvered timber ventilation lanterns, which provide a passively
ventilated space beneath. This mitigates the need for mechanical ventilation and its associated cost, and environmental
impact. Every aspect of this project, from the wide community
consultation, to the teamwork of the conservation, design and
contracting teams, ensured the success of this conservation
project.

Carrickfergus Castle is a reminder of Ireland's masonry
culture, which still dominates Irish construction. In the construction of the twelfth-century castle, Irish stone masons
worked under the direction of craftsmen brought over from
France by the Anglo-Normans. *Vessel*, on the other hand, is an
all-Irish production from concept to exhibition; from sourcing
and harvesting in a Coillte forest to processing, precision cutting, woodworking and assembly in Longford by Glennon
Brothers Sawmills and Gem Joinery. When installed in 2013, it
was surrounded by the bare brick walls and columns of the Arsenale, once a massive production complex in Venice during
the pre-industrial era.

While wood is combined with stone and slate in Carrickfergus as a utilitarian but innovative expression, all materials have
their place in a functional building. *Vessel* requires a more artistic expression that requires the wood to work with, rather
than against, the brick surround in its new, albeit temporary,
location. The connection between the two materials is made
by stacking the work in 5,500 spruce sections, that mirror the

The principal trusses of the new roof of Carrickfergus Castle were constructed from wind-blown Irish oak. The ridges of both hipped roofs are topped with louvered timber ventilation lanterns providing a passively ventilated space beneath.

bricks of the Arsenale which, in their time, would have been cast in wooden moulds before each individual brick was hand-thrown from its casing. This was the starting point of what is, essentially, a meditative and spiritual piece, inspired by the Seamus Heaney poem "Lightenings viii". The poem refers to the Celtic-Christian legend as the monks of Clonmacnoise monastery find their prayers interrupted when "A ship appeared above them in the air". Unlike the miraculous airborne ship that made its way to the monastic site, *Vessel* made its way to Venice by water but the symbolic reference to Heaney's "out of the marvelous" experience is well made.

Vessel, *installation at Venice Biennale, 2013 by O'Donnell + Tuomey Architects. Sitka spruce from Coillte forests, processed by Glennon Brothers Sawmills and manufactured by Gem Joinery, is the only species used in the 9-m tall structure.*

The Brendan currach, which featured homegrown ash in the frame and two masts.

This symbolism also resonates with another vessel project which also involved Glennon Brothers Sawmills in the 1970s when Tim Severin, explorer and writer, sought out Paddy Glennon to provide specially cut and treated timber for a large currach to replicate an early Christian cross-Atlantic voyage. Severin needed ash for the frame and the two masts of his 11-m currach, which he planned to sail from Ireland to North America for his epic Brendan Voyage. The adventure was to prove that St. Brendan could have crossed the north Atlantic Ocean in a currach as early as the sixth century, long before the Norsemen or Columbus. Severin achieved this feat in 1977 without resorting to technology or modern equipment.

The restoration of Carrickfergus Castle, the creation of *Vessel* and the re-creation of the Brendan currach are important projects in the story of Irish wood as they create their own wood culture. It is a wood culture receptive to external and internal influences and skills, as well as native and introduced tree species.

Sometimes to gain acceptance as a material, wood has to make bravura statements like *Vessel* or the Beyond the Trees experience in Avondale. Both are bold visual declarations that demonstrate wood innovation and conservation at its finest.

The viewing tower at the end of the Tree Top Walk, Avondale Forest Park provides spectacular panoramic views of the Avondale and Co. Wicklow landscape.

Beyond the Trees is located in a forest so it makes the connection between the wood and the trees in what is essentially a recreational experience. Geared towards a younger audience this is both an educational and functional sculpture. Like *Vessel*,

Tree Top Walk in Avondale Forest Park's Beyond the Trees experience. The walk and tower combine children's play areas and forest flora and fauna interpretive points.

it is made from home-grown sawn timber, but in this case it is engineered for strength. It charts a journey along the treetops ending at another massive vessel or viewing tower almost 40m tall. This provides a panoramic view of the forest and parkland below where the story of contemporary Irish forestry began in 1904, when the State purchased Avondale House and surrounding lands to provide tree species' trials and a forestry school. That the timber used and the diverse tree species are predominantly introduced is a demonstration of the diversity of Irish forestry which has been embraced in the story of trees and wood.

While the wood for the tower and treetop walk is home-grown, the design and construction required a mixture of Irish and overseas expertise, knowledge and skills. The interaction between Irish and overseas designers and woodworkers is essential in creating a wood culture in Ireland. This interchange is evident in projects throughout the country, from Joseph Walsh's innovative furniture designs to boat restoration in Kilrush. It will also be required in mass timber-building design if Ireland is serious about decarbonising the construction industry.

Overleaf: Rain Bridge, Peitian Village, Fujian Province, China. Designed by Donn Holohan with Hong Kong University students and engineers, while local carpenters worked on the bridge construction.

Irish architects and engineers who work overseas are returning to incorporate best international practice in timber construction and design. This is evident in projects submitted to Wood Awards Ireland, a competition, which encourages Irish practitioners working abroad to enter projects.

The results to date are impressive. For example, Irish architect Donn Holohan, winner of the international category in Wood Awards Ireland 2016, demonstrates that Irish architects and designers are now not only working overseas, but taking the lead in global wood projects even in countries with strong wood cultures.

Holohan, based in Hong Kong and Ireland, explores emerging technology in wood design but is also at ease in traditional woodworking. In his award winning *Rain Bridge* he led a Hong Kong University project, which was a collaboration between architecture students, university engineers, carpenters, woodworkers and the community of Peitian Village in Fujian Province, where the bridge was erected. The covered bridge is an interlocking timber structure that draws on the long tradition of wooden buildings native to the region. "Each of the bridge's 265 elements is unique and integral, assembled under the supervision of traditional carpenters, who number some of the few remaining exponents of their craft," explains Holohan. Like all good design, the bridge combines the traditional and contemporary but with sustainability at the heart of the project. Holohan says:

Rain Bridge, Peitian Village, Fujian Province, China, designed by Donn Holohan served as a functional river crossing and a social space for locals to meet and trade their produce.

> This project seeks to offer an alternative mode of community redevelopment that references local crafts and traditions, and utilises sustainable materials and methods, to create both a social and physical infrastructure. Critical to the process was the integration of parametric digital design methodologies with more traditional construction techniques.

Projects as diverse as the *Rain Bridge, Vessel*, Beyond the Trees and the Carrickfergus roof restoration demonstrate the importance of collaboration between those involved in the design and making, and those who work the forests and the wider community where the works are based and where the raw material is sourced. This is the beauty of wood as it can be sourced within or close to the communities that benefit from its end use. Sometimes we forget this and we export the finished product while not realising that its best and most sustainable application lies in its use within its own place.

This was brought home forcibly by a group of Irish and Estonian architects in the project Wood Works based on the theme "There is a forest in my backyard, but my house is built from trees grown far away" (*Mu maja pole puudest, mis kasvavad kodu taga metsas*). This collaboration was initiated in 2020 and brought to fruition in 2022 by curators at Alder Architects, Ireland and b210 architects, Estonia. Despite the COVID-19 pandemic, the project included a diverse range of temporary exhibitions in wood and seminars emphasising sustainable timber sourcing in construction and design, and exploring wood's afterlife in reuse and recycling.

Building from locally sourced wood was the unifying aspect of all the Wood Work exhibitions, not least in *Butterfly Building*, a partnership between Robert Bourke Architects (RBA), Dublin

"Conversation Piece" by Hannigan Cooke Architects, Limerick and Ruu-miringlus Practice, Tallinn, Estonia.

and Creatomus Solutions (CS), Tallinn. The CS team are admirers of the economist Kate Raworth, who is a proponent of the circular economy with the emphasis on reusing, recycling and repairing wood rather than the linear economy that favours disposal. Wood lends itself to this approach as illustrated by RBA who sourced local timber for the main structure as well as recycled materials for insulation and reused timber elsewhere. Traditional usage fused with the contemporary as the installation also contained a digital appliance to measure the carbon footprint during construction. Reuse was also the theme of *Conversation Piece*, a collaboration between the Ruu-miringlus practice and Hannigan Cooke Architects. The aim was to develop "a design tool for existing and pre-used materials" through a year-long conversation of webinars during the COVID-19 pandemic. Both practices decided to forego conventional architectural drawings and create an educational space which features information on wood as a renewable resource.

Although little more than half the size of the island of Ireland, Estonia has 2.5 million ha of forests, more than three times the size of the Irish forest estate – north and south – so Estonian architects are much more likely to have a forest in their back yards than their Irish counterparts. Yet, despite different forest and wood cultures, this collaboration worked creatively and functionally as the five Irish and five Estonian practices provided their own distinct visual language that never strayed from the core project objective of sustainability.

Both countries share a number of native tree species including birch, Scots pine, alder and common ash. The most important native species in Estonia is the native Norway spruce which although not native to Ireland, grows well here, but the preference is for Sitka spruce. Birch is the species that both countries share in terms of scale but not in use. While Ireland has 47,000ha of birch, Estonia has some 750,000ha of the species. As a result Estonia has created a veneer industry around birch while it is mainly used for firewood in Ireland but with the introduction of "improved" birch species, it will eventually have added-value uses.

In the Wood Works exhibition Peter Pere Architects and Workshop Architects combined to use birch in the round with plywood to create *Raw Potential*. While the two teams weren't advocating using round or raw timber for all construction, they maintained that architects and specifiers should "consider to what degree material needs to be processed to leave the greatest potential for reuse".

Common or European ash is a species that is native to Ireland and Estonia. These projects serve as a bitter-sweet reminder about the vulnerability of ash which is being infected with the deadly ash dieback disease. This disease is now common to Ireland and Europe, including Estonia. And ash is not the only species that is vulnerable to damage from insect, fungal and other causes including windthrow and fire, which are often

"Butterfly Building" by Robert Bourke Architects, Dublin with Creatomus Solutions, Tallinn.

climate-related. Larch, Norway spruce and chestnut are only a few species now threatened just as elm faced similar damage in the past which virtually wiped out the species. The advantage of trees and wood over fossil-based material lies in their renewability, which is assured providing trees are healthy and disease-free.

In an open economy like Ireland, our island status is no longer a protection against the introduction of tree diseases, which is why plant protection measures are vital to safeguard both the wood and the trees. How we nurture this great resource is reflected in our relationship with the tree and wood cycle, as Seamus Heaney outlined, "No bit of the natural world is more valuable or more vulnerable than the tree bit".

CHAPTER SEVEN

The Right Trees in the Right Places for the Right Reasons

I planted in February
A bronze-leafed beech,
In the chill brown soil
I spread out its silken fibres.
[...]
It is August now, I have hoped,
But I hope no more —
My beech tree will never hide sparrows
From hungry hawks.
Patrick Kavanagh, from "Beech Tree", 1936

I walked under the autumned poplars that my father planted
On a day in April when I was a child
Running beside the heap of suckers
From which he picked the straightest, most promising.
Patrick Kavanagh, from "Poplar Memory", 1937

Our relationship with trees is defined by our ability to protect and enhance the forests and woodlands we have inherited and to create new forests where they are needed. The choice of tree is limited by the soil and site conditions where it is planted, while the reasons for planting are limited only by our own imagination.

But function and imagination are interlinked in tree planting and care as the poet Patrick Kavanagh discovers. His expectation that his beech tree will "hide sparrows / From hungry hawks" never materialises, despite careful planting as he "spread[s] out its silken fibres" and "[fixes] it firm against / The worrying wind". Thus, he learns the first lesson in tree husbandry and silviculture, which is to ensure that species and soil conditions are compatible. Beech prefers free-draining mineral soils rather than the wet Drumlin soils of Kavanagh's Co. Monaghan landscape.

The soils around his home place of Mucker near Inniskeen are more conducive to poplar growing and Kavanagh is aware of this, at least poetically, if not silviculturally. His father's selection of poplar trees, unlike his son's choice of beech, is based on imagination but is tempered with the limitations of site and the availability of cuttings from existing trees. His selection of "the straightest, most promising" suckers for planting allows him to part-fulfil his dream: "[Of] poplar forests in the waste places / And on the banks of drains."

The son's romantic hope of hiding sparrows from hawks contrasts with the father's more practical objectives which, although not stated, would have been to provide shelter and possibly firewood. What unites father and son is their singular purpose; to add beauty to the largely treeless landscape of Co. Monaghan, still one of the least forested counties in Ireland. Both achieved their personal aesthetic: the father in the actual; the son in the imaginary.

The experiences of Patrick Kavanagh and his father illustrate the need for an understanding of which trees to plant and where, if the expectation of the grower is to be realised. The ultimate message is that tree species selection – the right trees – and function – the right reasons – are an intrinsic and inseparable part of woodland creation.

The right trees in the right places

Come then, and learn what tilth to each belongs
According to their kinds, ye husbandmen…
Virgil, *The Georgics* by Virgil, c29BC, translated by James Rhodes, 1881.

Selecting the right trees in the right places has been the forester's guiding principle since forests have been managed. Over 2,000 years ago, Virgil applied the principle to agriculture, silviculture and viticulture in *The Georgics*. It applies both to reforestation, where forests need to be replenished and afforestation, where new forests and woodlands are created or recreated.

Reforestation or restocking provides an opportunity to maintain the forest continuum in the presence of existing trees and woodlands, while afforestation presents the greater

The Coronation Plantation near the Sally Gap in Co. Wicklow, planted in 1831 deserves restoration.

challenge of establishing future forests in a landscape where there is an absence of trees and an absence of a forest and wood culture. Nature dictates the species choice available to the forester on reforestation sites unless it is necessary to introduce alternative tree species. where the previous crop has failed due to disease.

This is true especially in countries and regions, fortunate to have largely unbroken forest tradition. Thus, forests in western North America maintain a constant tree species balance down through the millennia comprising predominantly Douglas fir, supported by a diverse range of mainly coniferous native species to replenish forests that have been harvested or destroyed by fire or other agents. These conifers include western hemlock and western red cedar as well as spruce, pine and fir species. While not as widespread, native broadleaves including alder, birch, aspen, ash and maple are also present. In Scandinavia there is heavy reliance on just two species – Norway spruce and Scots pine – with few broadleaves apart from birch. Further south, in central and western Europe, there is a more even conifer-broadleaf balance. Although there is still a high dependence on Scots pine and Norway spruce, here there is much greater broadleaf diversity, including beech and oak.

Forestry and agriculture, near Killavullen and Nagle Mountains, Co. Cork. The forestry option needs to be presented in terms that farmers relate to and understand; that it is a crop, which complements rather than threatens farming.

Unlike Irish foresters, northern hemisphere foresters have a much more extensive palette of indigenous species to choose from when planting trees. Apart from Scots pine, all coniferous tree species growing widely in Europe are not native to Ireland. All European spruce, larch, fir and pine species (with the exception of Scots pine) are non-native. Broadleaves, including beech, sycamore, sweet chestnut, field maple, Norway maple and lime failed to colonise Ireland as natives. You could count on one hand the number of native broadleaves that would provide a viable ecological and productive woodland: oak pure, certainly on good sheltered soils; cherry in mixture, probably; alder and birch maybe. Unfortunately that's it, especially now as ash and elm are no longer planted due to disease vulnerability.

Fortunately, all these European species grow well in Ireland, while many of the broadleaves are now naturalised. Even better performances are being achieved by species that originate in western North America where growing and maritime conditions are not dissimilar to Ireland.

Thus, foresters charged with reducing Ireland's dependence on imported timber at the beginning of the last century had only three native "timber" trees at their disposal – Scots pine, oak and ash – and a few more/others with limited potential as woodland trees such as cherry, alder and birch. The challenge to create viable forests was compounded as these species require good-quality land which wasn't available to foresters for most of the twentieth century. Unlike Europe, forestry was – and still remains in some parts of Ireland – the land use of last resort.

The mistaken belief that trees would grow on poor sites lasted for much of the nineteenth century in Ireland. In the absence of good silvicultural advice, two disastrous forestry projects finally proved that while trees don't require good agricultural land, they *do* need suitable soils, sites and climatic conditions to thrive. The ill-fated Coronation Plantation near the Sally Gap in Co. Wicklow, planted in 1831, provides compelling evidence of the need to match the right trees to the right places. Unfortunately it took a further misjudged project in Knockboy, near Carna, Co. Galway in the 1880s to finally illustrate that native and European species would neither thrive, let alone survive, on these inhospitable sites.

While it is true that most of Ireland was once under forest, the belief that Ireland's afforestation programme should somehow replicate native coniferous forest trees ignores the changes in soils, climate and land use that accompanied the decline of the primeval forests over the millennia.

For example, changing land use and climate contributed to the spread of raised and blanket bogs which coincided with the decline of native woodlands more than 7,000 years ago. Even the resilient pine – now Scots pine – Ireland's only native conifer forest tree, succumbed to the encroaching peat to such an extent that it was almost wiped out by the early Christian period. The well intentioned but inexperienced people involved in the Carna experiment would have had to dig down into three metres of wet peat to find evidence of the impressive ancient forest. The oak dugout canoe in the National Museum of Ireland, excavated from Lurgan Bog, Addergoole, Co. Galway is also evidence of the magnificence of the primeval forest. But that was before the oak forests succumbed to the encroaching peat. The acidic layers of peat that cover these ancient forests have created their own unique, but treeless ecosystem.

In the barren early twentieth-century landscape, Irish foresters were faced with the unenviable task of establishing a productive forest resource. The scant evidence of species that could adapt to Irish conditions was limited mainly to old woodland estates as a result of an afforestation scheme administered by the Dublin Society – now the Royal Dublin Society (RDS). Initiated in 1745, it lasted until the early nineteenth century after the Act of Union. This unique State-private partnership – the first of its kind in the world – began a trend of planting mainly European species which continued into the first few decades of the last century, but then species selection changed dramatically as the land made available for afforestation changed. Most of this bare land was a mixture of uplands, bogs and nutrient-deficient exposed sites, not capable of supporting European or Irish species.

Conscious of this, foresters turned their gaze to the northwest of another continent with a similar latitude and temperate maritime climate to Ireland. Western North America was singled out as the most suitable seed source, based on the performance of trees introduced to Ireland from this region in the 1830s - some of which still survive. The botanist and forester Augustine Henry was convinced that trees "native of the Pacific slope of North America [would] find in Ireland a climate exactly like that in which they occur at home".

During the years following the purchase by the State of Avondale near Rathdrum, Co. Wicklow in 1904, trial tree plots were established in the estate. In all, 100 different tree species from around the world were planted to determine their potential in Irish conditions. The results of these trials influence tree species selection to this day. The impressive performance of species native to western North America proved Henry correct. Sitka spruce, Douglas fir, lodgepole pine, western hemlock and western red cedar were introduced into the canon of Irish species. As a result, foresters whose tenet is to match trees to suitable sites and climatic conditions, planted these species to maximise timber production on the poor land made available for forestry up until the 1990s.

In this regard, foresters achieved the main objectives set by the State, which were social and economic. They established a resource, mainly on inhospitable landscapes on the poorest of soils, which was capable of dramatically reducing the nation's dependence on softwood timber imports and providing an alternative use for unproductive land, while creating badly needed, sustainable employment in rural areas.

Within these objectives, planting the right trees in the right places proved correct for State forestry. But when the State withdrew from afforestation in the 1990s, a new forest paradigm was introduced. Within a few years after the State-owned company, Coillte, was established in 1988, public afforestation was replaced by private – mainly farmer – planting. This dramatically changed the scale, ownership, species selection and objectives of the national forestry programme.

The right trees for the right reasons

> ... the barren woods
> That crown the scalp of Caucasus, even these,
> Which furious blasts for ever rive and rend,
> Yield various wealth, pine-logs that serve for ships,
> Cedar and cypress for the homes of men;
> Hence, too, the farmers shave their wheel-spokes, hence
> Drums for their wains, and curved boat-keels fit...
> Virgil, *The Georgics*, c29BC, translated by James Rhodes, 1881.

When Virgil surveyed the devastated landscape of northern Italy in the final stages of a brutal civil war, he saw ruined farms and a depopulated countryside. *The Georgics* began to take shape in his mind probably around 37BC; restoring the damaged land to productive use is a central theme of his didactic poem. Virgil allows the farmers, who work the land, to have the final say in its nurturing and raising crops in the forest, farm or winery and to reap the benefits of their labour.

Over 2,000 years later, the State would look to farmers in Ireland to continue the work of restoring the lost forest resource which it began in 1904. Farm forestry has introduced a different set of challenges in establishing forests, as these are smaller in scale than State forests and are also located on better-quality land. As a result, there are greater opportunities for tree species diversity, small-scale woodland development and agroforestry, as well as integrating forestry and farming, and optimising tourism and other non-wood benefits.

Even before the shift from public to private afforestation, foresters had been placing greater emphasis on the non-wood aspects of forests including recreation, climate change mitigation, food foraging and biodiversity. These management objectives are not new as forests have long been regarded as environments for non-wood activities. For example, the State's visionary decision to create an open forest policy in the 1970s was an acknowledgement that Ireland's forests are a multipurpose resource whose value is measured, not just in economic and yield terms, but in their total contribution to society. That these forests contain 62.5% commercial conifers hasn't dampened the enthusiasm of walkers who make 29 million visits annually to Ireland's recreation forests and woodlands in the south – mainly to Coillte forests – and nine million in the north where the conifer cover is even higher.

Despite the success of conifers as versatile fast-growing species, the forestry programme has become too one-dimensional because of its heavy reliance on Sitka spruce. The risk of disease to this marquee tree has placed greater emphasis on species diversity, which has been a feature of Irish forestry since 2000. Species composition has broadened as planting location changed to good, albeit marginal agricultural land from exposed sites including blanket bogs where afforestation is no longer carried out.

Broadleaf planting, which amounted to 5% of total planting for most of the last century, increased to an average of 32% in the 10 years up to 2012, when the ash dieback fungal disease caused farmers to rethink their broadleaf planting strategy.

State funded afforestation now requires a minimum area of 35% broadleaves and open spaces. The grant condition, strongly favour native and naturalised broadleaf afforestation. In the case of species such as oak and beech, this requires

Integration of forestry and agriculture, Glanmore and Devil's Glen, Ashford, Co. Wicklow with broadleaves on the lowlands and mainly conifers – Japanese larch, Sitka spruce, Scots pine and western hemlock – interspersed with beech and birch on the poorer soils and more exposed uplands.

continuity of management over a 100-year rotation compared with 30 to 40 years for conifers.

For most farmers, forestry provides, inter alia, a pension fund. Planting a sizable proportion in broadleaves begs the question: Can farmers be expected to establish and manage a crop that will at best provide a return for their grandchildren? Surprisingly, farmers' planting programmes so far this century prove that they are willing to plant significant numbers of mainly native broadleaves. However, they are adamant that they need to establish medium-rotation tree crops to support or carry the longer rotation broadleaf crops for future generations.

Conifers are viewed by foresters and farmers as the enablers to plant broadleaves. Without this species mix option, the danger is that many private landowners, especially farmers, won't consider forestry as a viable long-term land use option. Forestry also differs from other land uses: once the forest is established, it must be retained in perpetuity due to the replanting obligation enshrined in the Forestry Act 2014.

While private investors are showing an interest in afforestation, most of the forests of the future will be established and managed by farmers or by foresters and contractors on behalf of farmers. Farmer ownership presents major challenges in Ireland as there is little experience of farm forestry, unlike Europe. If Ireland is to achieve the government's target of 18% forest cover, the forestry option needs to be presented in terms that farmers relate to and understand; that it is a crop, which complements rather than threatens farming.

Notwithstanding the benefits of tree species diversity, farmers view forestry as a land use and income source that supplements earnings from agriculture. So, for a tillage farmer with a forest on a less productive part of the farm, maximising wood production is as natural as maximising malting barley yield. There isn't a conflict of interest between the forest and agriculture objectives. The forest, through good silvicultural practice, maximises yield to provide quality timber for construction, furniture, fencing and other uses. Tillage – depending on the crop – maximises yield for feedstuffs to the livestock sector or the raw material for industries such as malting, milling, cereals and other food products. In the future, end uses may even merge, as energy and biofuels are developed in timber and some tillage crops.

The role of forestry as a marketable crop to support downstream wood industries and employment is likely to remain a major objective, but it is not the only one. While productive forestry is likely to continue to be the cornerstone of forestry policy, there is little doubt that multipurpose – or non-wood – forestry will receive far greater prominence in the future. Multipurpose forestry includes leisure, tourism, water quality, flood prevention, biodiversity and climate change mitigation.

The debate on the role of forestry in climate change mitigation has focused on forests as carbon sinks and providers of raw material for generating renewable energy. However, the combined role of forests and wood in sustainable construction, wood energy and other applications to mitigate climate change is a key aspect of the climate change debate. As a result the connection between carbon sequestration in the forests and carbon storage in wood products, especially in construction, is often missed by many commentators. This applies to timber frame house construction, while engineered softwood timber has already transformed sustainable building and sustainable living in Europe and Canada. The potential of engineered wood in medium-to-high-rise buildings is a key aspect of the role of wood in the bioeconomy, not just in relation to carbon storage during the life of the building but also because of its potential in displacing fossil-based materials as outlined in the previous chapter.

CHAPTER EIGHT

What Trees? Species That Adapt to Irish Soil, Site and Climatic Conditions

As treeless as Portugal we'll be soon, says John Wyse, or
Heligoland with its one tree if something is not done to reafforest
the land. Larches, firs, all the trees of the conifer family are going fast.
James Joyce, from *Ulysses*, 1922

James Joyce was well aware of the perilous nature of Irish forestry when writing *Ulysses*. The novel is set in 1904, the year after a devastating storm wreaked havoc on Ireland's few remaining woodlands. It was also the year Avondale estate was purchased as a State forestry training centre and experimental forest. It marked the first real discussions on how best to restore Ireland's forests, which had decreased to little over 1% of the land area. So, the arboreal decision-makers had almost a blank canvas to reforest the country. The main influencer, botanist and forester, Augustine Henry, was aware that Ireland's moist, mild and maritime climate provides the ideal tree-growing environment to support a wide range of tree species, regardless of origin. More than a century later, the wide variety of tree species choice is reflected in the Forest Service grant aid programme which funds over 40 species in Ireland's afforestation programme (Table 8.1). These comprise coniferous and broadleaf productive species as well as "minor" conifers and broadleaves. Not all have commercial potential but they all have their own unique ecological and aesthetic qualities.

Because of Ireland's limited range of indigenous tree species, there is a high dependence on non-native naturalised and exotic tree species. Although the species menu is generous, growers should tread carefully when selecting the right trees, as the decision will dictate the quality of their woodlands for generations, along with wood and non-wood outlets for their produce. The species in the following pages provide growers, processors, manufacturers and end user with an array of possibilities, but the forest owner's choice depends on factors such as site requirements, soil types, climate and future markets.

The choice of trees will also depend on the landowner's objectives. It should be remembered that the forest owners cannot opt for a sole commercial objective because there is a mandatory requirement that caps commercial species at 65% of all afforestation sites with the remainder comprising broadleaves and open biodiverse areas. This condition provides the owner with the following five forest types at the afforestation or reforestation phase:

- Mixed forest with the emphasis on maximising commercial crop comprising 65% conifers and 35% native species and open biodiverse unplanted areas.
- Mixed woodland with the emphasis on long-rotation native or naturalised broadleaves but allowing some commercial short-rotation forestry, regardless of species selection.
- Native woodland comprising native broadleaves and Scots pine.
- Continuous cover forestry (CCF) whereby the forest is managed according to the principles of continuous cover or close to nature forestry.
- Agroforestry – integrating forestry with agriculture.

If a return on investment within a generation is the sole objective, then conifers tick all the boxes, including early revenue on a wide range of sites with tried and trusted markets for the end product. Sitka spruce, Norway spruce and Douglas fir have major market appeal, while Scots pine produces lower yields than exotic conifers, but reconnects with the ancient pre-Christian forests which qualifies it for grant aid under Ireland's Native Woodland Scheme. Scots pine also produces high-quality wood.

The increase in broadleaf planting in recent years demonstrates a desire for mixed-species forests, which is now being achieved in planting programmes. Unfortunately, ash, which is the most versatile broadleaf, is unlikely to be grant-aided in the foreseeable future while ash dieback disease caused by the fungus *Hymenoscyphus fraxineus* remains a threat to the species.

Farmers who wish to combine forestry and farming should consider the State agroforestry initiative. This underutilised scheme integrates well with agriculture, especially grazing, silage and hay production. It is ideal for first-time woodland entrants as the scheme is grant-aided for areas as small as a half hectare, or less than an acre. The scheme is suitable for native (oak and cherry) and naturalised (sycamore and sweet chestnut) broadleaves, while fruit and nut-producing species are also considered.

Ultimately the choice of species and management regime lies with the landowner. But the safest, and most satisfying approach, is to select at least two species including one main commercial crop species and one or more long rotation broadleaf – native or naturalised – species; conifers for a relatively fast return on investment and broadleaves for the distant future and to reconnect with the lost primeval forest.

ACHIEVING A NATIVE, EXOTIC

Prejudice should not exclude foreign trees; the question is whether they will grow well as forest trees.
Augustine Henry

When the 2023 Forestry Strategy proposed an afforestation programme of at least 50% native species, it did so without any scientific or research analysis. The strategy contains no assesment of national land availability for forestry and its suitability for growing quality native species. The ecological reasoning is also unsound as it is devoid of analysis of soil types to support a major native tree afforestation programme.

The strategy references Ireland's Climate Action Plan which calls for species that will displace fossil-based material in construction to help Ireland achieve net zero by 2050. Conifers will play a major role in cement, brick and steel substitution by mid-century and beyond, while broadleaves will play little or no role in displacing fossil-based material even by the next century. Conifers will also play a major role in renewable energy, just as they do in climate leading countries such as Sweden and Denmark.

Nobody disputes the argument that native species are needed and make for a more diverse forestry programme but apart from Coillte's pledge to achieve a 50% native tree afforestation programme by mid-century, there is no analysis of how private growers – mainly farmers – can achieve this ratio. In recent years, Ireland has achieved 40% native species afforestation programmes, which, is marginally greater than Scotland. However, Scotland has much greater planting programmes as coniferous afforestation is seen as an enabler to plant up to 4,000ha annually, compared with 800ha in Ireland. The strategy never asks the question why Ireland, unlike Europe, has been more reliant on introduced rather than native species.

ND NATURALISED BALANCE

The answer is simple: we have a dearth of native trees, shrubs and other vascular plant species. Unlike European foresters, who have up to ten native timber producing conifers to choose from, depending on the country, Ireland has one (Scots pine). Likewise European countries have eight quality native broadleaves to choose from while Ireland has two (oak and possibly cherry – alder and birch are secondary species in Ireland as they are in Europe, apart from Finland).

If, like Europe and the US, we had an abundance of native species, there would be no argument for not planting them. We are fortunate however, in having the climate and sites to grow the highest yielding forests in Europe.

Climate change is now forcing countries to increase forest cover and sustainably manage existing forests with native and non-native species. For example, some European foresters are looking to Douglas fir as a replacement to their damaged Norway spruce forests, while Portugal is experimenting with non-native red oak on former drought affected farmland.

Ireland's foresters need to diversify tree species including diversification of native, naturalised and introduced species with clear environmental, social and economic goals, supported by research. Foresters, farmers and other landowners have an innovative role to play in establishing more diverse forests. But the ultimate goal should be to produce quality forests and not let prejudice solely dictate tree species selection as Augustine Henry advised. We need to nurture the natives and the naturalised but also to promote the best of the exotic species that have a solid track record in maximising decarbonisation in the forest and in displacing fossil-based materials in construction and energy.

Table 8.1: *Grant–aided tree species in Ireland.*

SPECIES	BOTANIC NAME
WOODLAND AND FOREST TREES *	
CONIFERS	
Scots pine	*Pinus sylvestris*
Lawson cypress	*Chamaecyparis lawsoniana*
Leyland cypress	*Cupressocyparis leylandii*
Monterey cypress	*Cupressus macrocarpa*
Western hemlock	*Tsuga heterophylla*
European larch	*Larix decidua*
Japanese larch **	*Larix kaempferi*
Hybrid larch **	*Larix x eurolepis*
Douglas fir	*Pseudotsuga menziesii*
Grand fir	*Abies grandis*
Austrian pine	*Pinus nigra (var. nigra)*
Corsican pine	*Pinus nigra (var. maritima)*
Lodgepole pine (N. Coastal)	*Pinus contorta*
Lodgepole pine (S. Coastal)	*Pinus contorta*
Monterey pine	*Pinus radiata*
Norway spruce	*Picea abies*
Serbian spruce	*Picea omorika*
Sitka spruce	*Picea sitchensis*
Western red cedar	*Thuja plicata*
Coast redwood	*Sequoia sempervirens*
BROADLEAVES	
Common alder	*Alnus glutinosa*
Italian alder	*Alnus cordata*
Ash **	*Fraxinus excelsior*
Beech	*Fagus sylvatica*
Southern beech	*Nothofagus procera, N.obliqua*
Cherry	*Prunus avium*
Sweet chestnut	*Castanea sativa*
Lime	*Tilia cordata / T. platyphllos*
Norway maple	*Acer platanoides*
Sycamore	*Acer pseudoplatanus*
Pedunculate oak	*Quercus robur*
Sessile oak	*Quercus petraea*
Red oak	*Quercus rubra*
Downy birch	*Betula pubescens*
Silver birch	*Betula pendula*
SOME SPECIES SUITABLE FOR AGROFORESTRY***	
Oak	*Quercus robur, Q. petraea*
Cherry	*Prunus avium*
Sycamore	*Acer pseudoplatanus*
Sweet chestnut	*Castanea sativa*

KEY: Native   European and naturalised   Exotics – non European

*Other native tree and shrub species, may also be grant–aided. ** Grant aid suspended because of disease vulnerability. ***Other broadleaves and conifers considered on a site to site basis.
Source: Adapted by D. Magner from Department of Agriculture, Food and the Marine, 2020.

This approach to mixed species may be likened to the Chinese proverb: "If I have two coins, I will buy bread with one so that I may live and a rose with the other so that I may have a reason for living." In forestry terms, the owner may plant conifers to live and oak for a reason to live. The fundamental objective is to achieve a balanced, viable forestry programme with economic, social and environmental benefits.

If forestry is a multifunctional land use then it has to have multipurpose objectives, However, most farmers and forest owners who work the land are entitled to achieve a return on investment even if that income is well into the future, so choosing the right trees for the right markets is likely to be a major factor. And the markets for timber have remained consistent over the past century.

Markets are divided into softwood and hardwood. In Europe, Russia, western North America and much of China, softwoods such as Norway spruce, Scots pine and Douglas fir along

Mixtures of broadleaves and conifers enhance the landscape, such as here in Avoca Forest, Co. Wicklow. The forested landscape comprises predominantly introduced species including beech, larch and spruce alongside native Scots pine and occasional birch.

NATIVE, NATURALISED AND EXOTIC TREES IN MIXTURES

Categorising species into native, naturalised and exotic is a convenient exercise but overly simplistic as it implies that tree species have their own distinct territories and ecosystems. This is true to a degree, especially in relation to indigenous woodland habitat types, but various tree species are often found in mixtures, where native, naturalised and exotic species converge and integrate.

In such an environment, tree species mixtures encourage diversity in the forest tree composition but also in the understorey shrub and herb layers. While Irish forests have been occasionally criticised for their reliance on single species, once the forest inscape is explored, different species survive – and thrive – side by side, although there is always an element of competition for space.

The National Forest Inventory, 2017 found that "just over a quarter (25.9%)" of forests were "identified as monocultures" with almost half containing three or more species from a total of 53 different tree species. It found that nearly half (48.8%) of forests comprise at least three species while 18% contain five or more species.

with a wide range of spruce, larch and fir species dominate the market. Hardwoods such as oak, beech, maple and cherry still continue to be in demand for flooring, furniture and joinery. In addition, niche markets are available for a wide range of other softwoods and hardwoods as a new wave of innovative architects, engineers and designers turn to wood as a sustainable, twenty-first-century construction and design medium.

This selection is based on the criteria that they all grow well in Ireland and provide both wood and non-wood benefits. I have chosen 16 species that adapt well to Irish conditions comprising six natives, five European and naturalised species, and five non-European exotics.

THE NATIVES

Tree species that arrived in Ireland naturally after the last Ice Age are classed as native. While Ireland has a wide variety of tree and shrub species, most were introduced. Ireland has no more than 30 indigenous native trees and shrubs, less than a third of some European countries and only a small fraction of numbers found in tropical forests.

While we have no more than five timber-producing trees of scale, the value of native woodlands extends beyond the commercial as they provide major biodiversity and heritage benefits. Forest policy in Ireland promotes the enhancement and conservation of existing native woodlands. It also promotes the establishment of new native woodlands either as stand-alone woods or as copses adjacent to streams, rivers and lakes to protect water quality as well as providing corridors for wildlife and biodiversity to connect with non-native forests. The benefits of this approach are recognised by the State, which is part-funding a Native Woodland Scheme to conserve existing native woodlands and increase native woodland establishment to 30% of all planting.

I have chosen five natives but forest owners may experiment with other species including holly, hazel, crab apple, aspen, willow, blackthorn, hawthorn and whitebeam. The following species provide the greatest multipurpose potential in the restoration of Ireland's native woodland resource: oak, Scots pine, ash, alder, birch and cherry.

Oak

Although the great oak forests of Ireland have long disappeared, it is still the main species in the few remaining semi-natural woodlands around the country. It also grows alongside other native and introduced species in mixed woodlands, while impressive oak specimen trees can be seen in the countryside including fields, parklands and occasionally interspersed in hedgerows. With the demise of ash, it is now the undisputed national tree.

There are two native species of oak: sessile oak (*Quercus*

petraea) or *Dair ghaelach* in Irish and pedunculate oak (*Q. robur*) or *Dair ghallda*. Identification is not difficult as the leaves have a unique round lobed shape. Clues to identifying Ireland's two native oaks are in their names: sessile, meaning stalkless, denotes an acorn without a stalk (below), while pedunculate has acorns with peduncles or stalks. The leaves are the other way round with stalked leaves on the sessile oak while the leaves of the pedunculate are stalkless. Both species can hybridise which makes identification less clear.

The 15-m oak canoe – dating to 2500BC – in the National Museum, demonstrates oak's once enormous size as a woodland tree, as discussed in Chapter 4. Because it was highly prized it was also vulnerable to over-exploitation. Oak was widely used down through the millennia in boat and shipbuilding, construction, tanning and charcoal production for iron-ore smelting furnaces. By the mid seventeenth-century, tree cover in Ireland was only 2.5% of the land area comprising mainly native broadleaves – there were practically no conifers and few non-native broadleaves in Ireland at this stage. We have little idea how much of the land area of Ireland was covered by oak in the seventeenth century as it competed with ash, birch, elm, alder and other natives. By the beginning of the last century when total forest cover had reduced to almost 1% of the land area, oak was competing with natives and introduced species planted from the eighteenth century so it represented less than the still paltry 0.3% of the land area it covers today.

Derrycunihy sessile oak woodland, Co. Kerry. Normally there is a diverse range of ground cover vegetation but this area has been overbrowsed by deer.

The acorns of sessile oak are stalkless, which is the opposite of pedunculate oak.

OAK – IRELAND'S NATIONAL TREE

Oak's prominence in the land-scape goes back to the primeval forests, when it covered large swathes of land. It was recorded as a Class I or 'chieftain' tree in the Brehon Law tree list. It features in poems from the Christian period onwards, including Mad Sweeney or *Suibhne Geilt* during the deranged king's seven years of wandering after the Battle of Moira in AD637. Sweeney referred to it – in J. G. O'Keeffe's translation – as "high beyond trees".

Oak features in some 1,300 place-names, from Dervock (*Dairbhóg* – little oak grove) in Co. Antrim to Valentia (*Dairbhe* – place of the oaks) in Co. Kerry. In between, its name appears in Edenderry (*Éadan Doire* – hillbrow of the oakwood) in Co. Offaly, Kildorrery (*Cill Dairbhre,* also *Cill Darire* or church of the oaks), Co. Cork and Derreen, (*An Doirín* or small oakwood) in a number of counties. It also appears in the less familiar name *omna* in place names such as Portumna (*Port Omna* or landing place of the oak), Co. Galway, which is why it features as the emblem for the local hurling club.

It re-establishes the link between the pre- and post-Christian period. A T Lucas and other scholars agree that Kildare (*Cill Dara* – [Saint Brigid's] church of the oaks) and Derry (*Doire* – oak wood, originally from *Doire Choluim Chille* or Saint Colmcille's oakwood) originated at a time when both saints established churches and monasteries in the sixth century.

We have however, sufficient native oak woodlands with the potential not only to conserve this major native resource, but to greatly expand its cover. It exists pure and in mixture in woodlands as diverse as Killarney, Glengarriff and Uragh in the south, Brackloon, St. John's Wood and Old Head in the west, Abbeyleix and Tullamore in the midlands and Glendalough, and the Vale of Clara in the east, while small but significant oak woodlands exist throughout Ireland. These are in urgent need of conservation, enhancement and expansion. They are valued as ecosystems, capable of carrying a wide range of related layers of flora, as described by John Cross and Kevin Collins:

> The shrub layer, which is typically dominated by holly, may be poorly developed due to overgrazing or may form dense stands, especially following a relaxation of grazing pressure. Holly is characteristically replaced by hazel on more fertile sites, where ash may also occur. A dwarf shrub layer of bilberry and sometimes ling heather is typically present. The herb layer is usually species-poor and often dominated by woodrush with abundant ferns, e.g. hard fern, common polypody and bracken. Species such as bluebell occur where the soil is more fertile, e.g. beside streams or at the base of slopes. Many of these woodlands, particularly in the west of Ireland and in sheltered humid sites elsewhere, are noted for the richness and luxuriance of mosses, liverworts and lichens.

Oak has been enjoying a slow revival, especially on suitable sites, since the introduction of the Native Woodland Scheme and the Forest Environment Protection Scheme by the Forest Service, while a new small-scale native woodland scheme allows farmers and other landowners to plant up to 2ha without the need for a licence while providing attractive grants and tax-free annual premiums.

Although oak prospers in good mineral soils, it is much more adaptable than it is given credit for and will grow in a wide range of site conditions. However, like most broadleaves, it grows best on fertile or moderately fertile low elevation, free-draining sheltered sites. Pedunculate oak favours moist brown earths and heavy clays that allow deep rooting, although its tap root will penetrate compact stony soils as well as gley soils over sandstone, but heavy gley soils over limestone glacial till, such as exists in parts of counties Leitrim, Cavan, Roscommon and neighbouring counties, should be avoided. Sessile oak, which is more adaptable than pedunculate, will tolerate lighter shallower soils in upland, more exposed sites. It favours acidic soils (pH 4.0-6.0) while pedunculate also grows well over a higher pH range (pH 4.5-7.5), performing well in neutral and slightly alkaline soils.

Oak responds to natural regeneration and coppicing but where planting is required, good strong nursery saplings are essential – preferably sourced from home-grown seed. Fertiliser application shouldn't be necessary on reasonably fertile soils. Oak needs to be carefully managed and regularly thinned throughout its rotation to produce straight, well-formed stems.

There is no discernible difference between the wood of pedunculate and sessile oak. Both produce hard and naturally durable heartwood, which varies in colour from yellowish brown to deep brown. Because of its wood structure, the faster oak grows, the stronger it is. It is a timber prized by woodworkers and architects as its end uses are diverse, including furniture, construction, joinery, flooring, turnery, exterior cladding and boat-building, while naturally or manually pruned oak provides high-quality veneer.

Although oak takes between 100 and 140 years to fully mature, its value can't be measured in narrow economic terms alone, although future generations will find high added-value commercial uses for the species. Because of its associated fauna and flora, it needs to be promoted and managed as a key ecosystem species in the re-establishment and enhancement of Ireland's native woodland heritage.

Scots pine

Scots pine is Ireland's only native conifer that combines commercial wood, heritage, aesthetic and other non-wood benefits. It has little opposition as there are only two other native conifers in Ireland – juniper and yew. Juniper is at best a bush, and while yew has considerable cultural, medicinal and specialised wood qualities, commercial timber production is not one of them.

Scots pine (*Pinus sylvestris*) appears under a variety of names including Scotch pine, fir deal and Scotch fir as in Seamus Heaney's poem "Crossings xxxi" when he observes windswept Scots pine that grow on the side of a road running through North Antrim bog: "Tall old fir trees line it on both sides./ Scotch firs, that is. Calligraphic shocks / Brushed and tufted in prevailing winds."

Its status as a truly native tree has suffered, mainly because of the widely held belief that it had died out by early Christian times and depended on introduced trees for its revival. This reasoning has never been fully explained. If it had died out during the early Christian period, why was it categorised as a "noble of the wood" in the Brehon laws which lasted up until the Norman invasion in the twelfth century? Claims of the extinction of native Scots pine seem to based, at best, on years of accumulated fiction characterised in the Irish proverb *Dúirt bean liom go ndúirt bean léi* …("A woman said to me that a woman said to her …") without any reference to fact or scientific research.

When comprehensive research was finally carried out on its ethnicity by Dr. Jenni Roche, Trinity College Dublin, her PhD thesis findings dismissed the premise that native Scots pine died out in the early Christian period (see panel). Based on vegetation surveys at eighteen pinewoods in Ireland and six in Scotland Dr. Roche stated: "The widely accepted hypothesis that *P. sylvestris* became extinct in Ireland has been rejected. "She explained the unbroken history of Scots pine in the study sites in Co. Clare:

SCOTS PINE ETHNICITY RECLAIMED

Research carried out by Dr Jenni Roche (below, with bog pine) for her PhD thesis at Trinity College Dublin settled the debate on the ethnicity of Scots pine. In a paper submitted to Irish Forestry (2019), the journal of the Society of Irish Foresters, she stated: "The widely accepted hypothesis that *Pinus sylvestris* became extinct in Ireland has been rejected."

Dr. Roche conducted vegetation surveys at 24 pine woods – 18 in Ireland and six in Scotland. "The pollen diagram from Rockforest Lough [Co. Clare] showed a continuously high *Pinus* pollen frequency (38-51% of total terrestrial pollen) from c.AD350 to the present," she reported. Macrofossil evidence demonstrated local presence of *P. sylvestris* around Rockforest Lough c.AD840. The available historical sources indicated a long history of woodland cover at Rockforest. Research extended to the nearby Aughrim Swamp showed a continuous *Pinus* signal from c.AD 350 based on peat core analysis. Her study should remove ambiguity about Scots pine's ethnicity. However, she urges caution about maintaining the species continuum and calls for further research and protection at the Rockforest site. "The [native Scots pine] population is of high conservation value but its rarity increases its extinction risk," she stated.

The pollen diagram from Rockforest Lough showed a continuously high *Pinus* pollen frequency (38-51% of total terrestrial pollen) from c. AD 350 to the present. Macrofossil evidence demonstrated local presence of *P. sylvestris* around Rockforest Lough c.AD840. The available historical sources indicated a long history of woodland cover at Rockforest. A separate analysis of a peat core from nearby Aughrim Swamp also showed a continuous *Pinus* signal from c.AD 350 to the present.

She makes positive claims for reintroduced Scots pine, which she also found in her research:

The Irish survey sites, excepting Rockforest, are thought to contain *P. sylvestris* of reintroduced origin. The vegetation of certain plots and groups resembled that of extant native pinewoods elsewhere or fossil assemblages from Ireland's ancient pinewoods. This suggests that reintroduced *P. sylvestris* woodlands are an important resource for Irish biodiversity, particularly given the country's low native woodland cover.

Remnants of the ancient Scots pine forest in Rockforest, Co.Clare.

THE RISE AND FALL OF SCOTS PINE

The literal Irish name for Scots pine is *An Péine Albanach* but other Irish names include *gíus*, *ochtach* and *ailm* which represents the letter "A" in the Ogham alphabet. The species colonised Ireland after the last glacial period some 10,000 years ago and covered large areas of the country for at least 4,000 years. Climate change leading to the formation of peat, coupled with burning and ground clearance for agricultural development are believed to be the main causes of its demise, beginning 6,000 years ago. Still, Scots pine continued for centuries as an important tree in medieval Irish culture. It features as one of the *airig fedo* or 'nobles of the wood' in *Bretha Comaithchesca* or "Laws of the Neighbourhood".

Scots pine almost died out during the early Christian period but experienced a mini revival from the eighteenth century and again during the period 1920 to 1951. During this time, it was a major tree in Irish forestry reaching one third of all planting programmes during peak years especially in the restocking of old woodland estates. It had a spectacular fall from grace in the late 1950s which has lasted up until recent times. It is now being promoted as a forest tree again, helped by attractive grants under Ireland's Native Woodland Scheme, but it is limited to 35% of the area, which suggest that the bias against this native conifer still exists.

Polestar, Letterkenny, Co. Donegal by Locky Morris. This sculpture demonstrates the industrial transmission pole usage of Scots pine and Douglas fir in connecting Ireland's electricity and telecommunication's network since the 1940s.

Opposite: Mature Scots pine stand in Dunanore Wood, Co. Wexford.

Dr. Roche's findings should remove ambiguity towards Scots pine as a native, although other factors may be at play in its rehabilitation as a major indigenous species. Because it is a conifer and an industrial timber producer, it sits uneasily in the narrative to restore Ireland's native woodlands which has been oversimplified to embrace only native broadleaves. This ambivalence is evidenced in *Management Guidelines for Ireland's Native Woodlands* by John Cross and Kevin Collins. In an otherwise excellent publication, the authors conclude with a set of management guidelines for native woodland but Scots pine, surprisingly, is omitted. Yet Scots pine is an acceptable "overstorey and major species" in the conservation element of the "Native Woodland Scheme" and anyone who visits native woodlands such as Glengarriff, Co. Cork and the Vale of Clara, Co. Wicklow will surely acknowledge its importance as a major native species. The authors highlight the role played by foresters and forest owners in the past in managing Scots pine "for wood production in a manner compatible with biodiversity conservation" which should be the aim of twenty-first-century foresters, forest owners and ecologists.

Scots pine has a wide natural distribution which includes most of Northern Europe, but its reach is wider. It is common in Eastern Europe and in parts of Russia but is found to a lesser extent in Southern Europe. Its western limit, once thought to be Portugal, is verified by Roche's research to be 8° 57' West at Rockforest, Co. Clare now.

Scots pine will grow on a wide variety of soils and is relatively easy to establish. Apart from alder and birch, it is the most adaptable of the native species but far superior as a timber tree. It grows well on soil types such as acid, brown earths and light, sandy non-alkaline soils. It also shows potential on low-nutrient sites including cutover midland bogs.

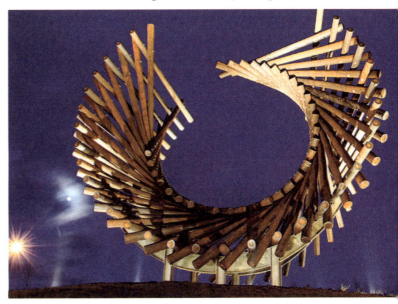

Cyril Hart in *Practical Forestry* maintained that "it can be grown almost anywhere including dry heath sites" where afforestation is no longer practised in Ireland – more's the pity.

Mature Scots pine is a magnificent specimen and forest tree producing excellent wood – red deal as it is known in the timber trade – suitable for a wide range of high-value end uses. It machines well and is used for joinery, furniture and flooring. If treated with preservatives, it has an extensive range of external uses including transmission poles, fencing stakes, posts, gates and railway sleepers.

Scots pine deserves re-evaluation even in sub-marginal uplands as in Scotland where it features in the once extensive Caledonian pine forests. Its distinctive red-brown bark and changing form as it matures make it an ideal tree in the lowlands and especially the uplands – its natural environment – where it once flourished. Despite its marginalisation in recent years, Scots pine is an integral part of the landscape, awaiting reinstatement as an important native ecological, commercial, and heritage tree in Irish forestry.

Scots pine, once the dominant species on the uplands could be restored in areas such as Maam, Co. Galway (above) where it was established under the Native Woodland Scheme alongside, birch, oak, mountain ash and alder. There is an opportunity to replicate the primeval forest just as Scottish foresters have been achieving in the restoration of the Caledonian Forest.

Ash

Save them, says the citizen, the giant ash of Galway and the chieftain elm of Kildare with a fortyfoot bole and an acre of foliage.
From *Ulysses*, James Joyce, 1922

The fictional citizen who appears in the "Cyclops" episode of James Joyce's *Ulysses* has many faults, but his plea to save the ash and elm species is both heartfelt and prescient. A half century after the publication of *Ulysses*, elm would face extinction while ash now faces a similar fate. Joyce based the citizen character on Michael Cusack, co-founder of the Gaelic Athletic Association. Ash would have been an important species for Cusack as it is used in hurley-making for Ireland's national game. Oak has been accepted as Ireland's national tree, but ash (*Fraxinus excelsior*) is a tree that is closer to the Irish consciousness in a number of respects. Ash is the wood of rhythm and action as hurley, oar, tool handle or in times past a " hand-weapon of a warrior" as depicted in J. G. O'Keeffe's translation of *Buile Suibhne* (*The Madness of Suibhne*). Its rootedness in Irish culture goes back to a time when it was revered as a *bile* or sacred tree, and to its emergence as Class 1 in the Old Irish tree list alongside the other nobles of the wood (*airgi fedo*), oak, holly, yew and Scots pine.

Oak's preference as Ireland's national tree is reflected in Irish placenames where it outnumbers ash in location and distribution. Ash has a more limited range, but derivatives of its Irish name – *fuinseog* – are common. While it appears in northern counties, it has strongest representation where it grows best, in Munster and Leinster, in places such as Lisnafunshin (*Lios na Fuinseann* – the ring-fort of the ash), Co. Kilkenny, Derrynafinchin, (*Doire na Fuinseann* – oak wood of the ash), Co. Cork and Killanafinch (*Cill na Fuinseann* – the church of the ash,

Distinctive fissured bark of mature ash growing in Townley Hall, Co. Louth, now threatened with ash dieback.

Co. Tipperary. The River Funshion (*Abhainn na Fuinseann* – ash river) typifies its favoured landscape as it flows along the border of counties Limerick and Tipperary before entering Co. Cork on its way to the Blackwater River near Fermoy. Ash place-names are also found in western and mid-western counties. In these counties it follows the ash growing – more nutrient–rich – eastern areas such as Cloonnafunshin (*Cluain na Fuinseann* – meadow of the ash and Funshin, (*Fuinseann* – place of ash), Co. Galway, Bearnafunshin (*Bearna na Fuinseann* – the gap of the ash), Co. Clare and Funshinaugh (*Fuinseannach* – place of ash), Co. Mayo. Farmers in particular embraced ash as a hedgerow species and when there was a resurgence of native tree planting from 2000 to 2012, they planted it extensively. They have a greater knowledge of ash than most other tree species as they know instinctively how it performs and interacts with its environment from watching it grow on hedgerows, near holy wells and in copses.

Overleaf: Magnus Celestii, *in laminated ash, Joseph Walsh Studios.*

Farmers planted ash in greater numbers than any other native species up to 2012 when average annual ash afforestation reached 840ha. Then, disaster struck. Ash dieback disease arrived with a vengeance in the guise of the deadly fungal pathogen *Hymenoscyphus fraxineus*, in 2012. Ash-planting ceased immediately while general afforestation declined. Confidence in forestry declined as the State refused to compensate ash plantation owners who watched helplessly as their crops were reduced to firewood use instead of high-value hurleys and furniture. Finally, in 2024, the State offered partial compensation, but it was too little, too late.

The destruction of ash is particularly apparent in Counties Tipperary, Meath, Kildare, Clare, Kilkenny, Waterford, east Galway and parts of Cork where the bulk of the new ash plantations were established. But its loss is noticeable throughout Ireland, as far north as a property in Ballykelly Forest, Co. Derry which is known locally as Caman Wood because of its long tradition of supplying hurley ash.

The loss of ash is incalculable from a commercial, heritage, social, sporting and even spiritual context. Just like elm, a few resistant trees will remain to remind us of what is lost and what may be restored if our researchers and scientists can develop a new generation of ash trees that resist dieback. Research continues in Europe and in Ireland – by Teagasc – to identify disease-resistant ash trees and build up a gene bank to eventually produce disease tolerant ash. In the meantime, hurley and ash furniture makers search for the last remaining healthy ash or look for alternative species.

Should Irish and European geneticists develop disease-resistant ash trees in future, it is likely that it would regain its former popularity as a woodland tree. On the right site, it will grow faster than any other native or European broadleaf and it will also outgrow most conifers. While ash grows on a wide variety of sites – even on poor soils – it will produce quality timber suitable for hurley manufacture and other high-value uses only on the richest soils, preferably in the lowlands where there is good shelter. It thrives on deep, fertile soils (limestone loams are ideal) which are moist but well drained.

As ash woodlands are associated with rich, moist soils, it's no coincidence that they have rich associated plant layers, including native tree and shrub species hazel, oak, birch, holly, blackthorn and whitethorn. The adaptability of ash to grow well in mixtures is not confined to native species but extends to naturalised and European trees including sycamore, beech and Norway spruce. It also supports a rich ground plant layer including meadowsweet, bluebell, nettle, anemone, primrose, violet, lesser celandine and dog's mercury.

The main lesson learned from the past disease outbreak is that ash will never be planted pure as a monoculture again. It will be established as mixed-species natural woodland to accommodate greater tree and flora biodiversity from the ground vegetation layer to a more varied tree canopy.

ASH WOOD – STRONG AND RESILIENT

Unlike slow–growing tree species that gain in strength with age, the faster ash grows, the stronger and more resilient it becomes. This is due to its wood structure. Within each annual ring there are two clearly defined rings – spring wood and summer wood. The spring – or early – wood has large pores while the denser layer of summer – or late – wood has tiny pores. Fast–growing ash produces a wide band of late wood which is denser and stronger than the early wood. This provides strong, resilient wood ideal for a range of products, especially hurleys – or hurls – for Ireland's national team sport, hurling (below).

Ash has a wide range of end uses including furniture, internal joinery, wood turnery, tool handles, picture frames, panelling and flooring.

Straight-grained and knot-free mature ash is well suited for veneer and plywood manufacture if manually or naturally pruned. Ash wood is perishable so is not suitable for outdoor use.

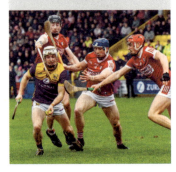

Birch wood is normally pale or light brown with occasional darker streaks or flecks.

Kenny McCauley, McCauley Wood-fuels and Marina Conway, Western Forestry Co-operative in a young birch plantation established on the McCauley family farm near Mohill, Co. Leitrim by Western Forestry Co-op. Birch adapts well on a wide range of soils from surface-water gleys to peaty gleys and from brown earths to iron pan podzols.

Birch

Downy birch (*Betula pubescens*) or *Beith Chlúmhach* in Irish and silver birch (*B. pendula*) or *Beith gheal* are both native to Ireland. They are difficult to tell apart. Mature downy birch tends to have cracked rather than the diamond-shaped patches of silver on its lower bark. The young twigs of downy birch have velvety white hairs without warts while the silver has hairless twigs with white warts. Charles Nelson claims the only way to really differentiate between the two is to study their "seeds and chromosomes". He opted for *Betula alba* – white birch – to cover both species as "all native birches belong to a single variable species".

Although a Class II tree in *Bretha Comaithchesca* or "Laws of the Neighbourhood" there is little doubt about its importance in the landscape. It has a wide distribution in placenames including Rossbehy (*Ros Beithigh* – birch wood), Co. Kerry, Glenveagh (*Gleann Bheatha* – glen of the birches), Co. Donegal, Ballybay (*Béal átha Beithe* – ford mouth of the birch), Co. Monaghan and Kilbehenny (*Coill Bheithne* – birch wood), Co. Limerick.

Downy birch is widespread in Ireland and grows well in a variety of soils and site conditions; from sheltered to exposed sites. Silver birch, not as common, prefers relatively good soils and tends to avoid exposed sites. Birch was one of the first pioneer species to emerge after the last Ice Age and paved the way for climax species such as oak, ash, elm and Scots pine. It is still regarded as a pioneer or nurse species, i.e. a tree that is the first coloniser on bare land capable of providing shelter and protection against frost as well as enriching the soil with leaf litter and decayed wood. These qualities benefit the quality woodland crop that follows. As a result, birch is seen as an accommodator rather than a major forest or timber tree in its own right, but this may be about to change. To dispel doubts about the quality and yield of both downy and silver birch, the results of a Teagasc tree improvement programmes which began in 2012 are impressive.

The performance of a downy birch research plot in the townland of Coolbaun, Co. Tipperary, a short distance from the shore of Lough Derg, has remarkable vigour and form. Planted in 1998, from 27 Irish provenances, this plot should provide commercial furniture lengths after two more thinnings or as early as 35 years of age. A nearby silver birch trial of mainly native with some German and French provenances is also promising. Birch now has assumed a new importance to enhance Ireland's limited menu of native tree species and is planted widely in native woodland scheme projects. It is an aesthetically important landscape tree because of its attractive bark and autumn leaf colour. It is comparable with beech, alder and ash in toughness and finish. It has potential for furniture-making and turnery although it is not durable so it is an indoor wood. It is a key veneer species in Finland where it is also valued for plywood production.

Alder

Common alder is a tree for the wetlands, where its ability to grow well even on poorly drained soils such as gleys is acknowledged by forest owners who are planting it in greater numbers in recent years. Its association with water extends to traditional uses such as sluice gates and water pipes, because it is durable when continually submerged. Much of the city of Venice is supported by alder piles. Although durable underwater, like most native hardwoods – except oak – it has poor durability above ground in outdoor use.

Common or black alder (*Alnus glutino*sa) appears as a Class II (commoner of the wood) in the Old Irish tree-list. Places such as Ferns (*Fearna*), Co. Wexford, Glenfarne (*Gleann-fearna*), Co. Leitrim, Gortnavern (*Gort na bhFearn*), Co. Donegal are derived from its Irish name – *fearnóg*.

Flowers develop on alder over winter as male – green to purplish brown – and female – red – catkins before maturing into dark-green cone-like fruits in the first year. These develop into winged seed the following season.

Alder has gained in popularity for a number of reasons. It grows well where other natives struggle, and qualifies as an approved species in the State's Native Woodland Scheme. It will adapt to a far greater range of soil types and site conditions than oak and has replaced non-native but naturalised broadleaves such as sycamore and beech, species that are more vulnerable to grey squirrel damage. Alder-dominated woodlands cover 19,700ha. There are few pure alder woodlands and it

Opposite: Michael Somers, Teagasc forestry advisor, inspects a downy birch research plot which was planted in 1998 in the townland of Coolbaun, Co. Tipperary. This is part of a national tree improvement programme which includes silver and downy birch.

tends to be associated with native willow or sally, ash and the introduced sycamore.

As explained in the COFORD publication *A Guide to Forest Tree Species Selection and Silviculture in Ireland*, by Ted Horgan and fellow authors, alder is a nitrogen-fixing tree: "Nodules attached to its roots, containing bacteria *(Frankie* species), enable the tree to make direct use of atmospheric nitrogen." Although planted pure in riparian zones, its main benefits may be a nurse tree, especially for oak on heavy soils. Compared with most broadleaves, it is a relatively short-lived species. Although there are some examples of long-lived alder in Ireland, these are exceptions.

It establishes easily and will achieve heights of up to 13m by year 15 but overall production over its rotation is not exceptionally high. According to the COFORD guide "it is such a short-lived species, it seldom grows to large sizes". The guide maintains that typical rotation lengths are between 30 and 50 years. Alder seems to be resistant to serious pest damage but there is evidence of crown dieback while *Phtophthora alni* is an increasing threat.

So, does alder deserve its status as a major native species in Ireland's afforestation programme? By 2010, it was assumed that it would be a major alternative species, especially for ash. Alder afforestation averaged 543ha over the following four years but from 2015 to 2023, average planting decreased to 86ha. On poor wet sites it has a role to play as it establishes quickly, so vegetation control is minimised, but improved birch may prove to be a better option.

However, if timber production is a major objective, tried and trusted coniferous species on poor sites are far more attractive options. These species provide superior yields and ready markets over similar rotations. Despite, its ability to produce attractive timber, the marketability of alder alongside other native timbers, in terms of future economies of scale and processing capability, are issues that have not been addressed satisfactorily in Ireland. However, a tree improvement programme being carried out by Teagasc is a positive development. This began in 2004 when a nationwide survey was conducted to identify healthy, vigorous good quality plantations. As a result an Alder Improvement Programme was initiated in 2005 with the aim of providing genetically improved alder tree germplasm for deployment in farm forestry.

Cherry

Wild cherry (*Prunus avium*) or *Crann Silíní Fiáin* or *gean* is probably the most attractive of all native trees because of its spectacular display of white blossoms in spring, its sweet-tasting edible fruit and attractive bark, while its wood is highly prized. Yet, it hardly features as a native species option in planting programmes. Cross and Collins give it only a cursory mention in *Ireland's Native Woodlands*. There is a question mark over cherry in a number of respects. It features as a Commoner of

When freshly cut, alder is pale in colour but darkens in the light to a reddish-brown colour, resembling cherry. It finishes well and can be stained and polished. It is a tree of the wetlands and can survive in flooded areas. Although the wood is non durable it resists decay when submerged in water even for centuries as demonstrated in Venice, where much of the city has been supported by submerged alder piles for centuries.

Common alder is a tree for the wetlands, given its ability to grow well even on poorly drained soils such as gleys, close to water and in partially flooded areas.

Cherry's heartwood can vary from rich red to reddish brown and darkens on exposure over time. It is ideal for furniture, as in MuTable by Knut Klimmek Furniture (below).

the Wood in *Bretha Comaithchesca* but its inclusion – as *idath* – is usually followed by an asterisk, questioning its authenticity (see page 32). It is rarely mentioned in early Irish poetry and it is missing in Sweeney's great roll call of native trees while its presence in folklore is sporadic. It is absent in placenames except for Cahernashilleeny (*Cathair na Silíní* – the stone fort of the cherries), Co. Galway, according to John Mc Loughlin.

Despite this, Padraic Joyce chose it as one of five important hardwood species in *Growing Broadleaves*, suitable for woodland establishment in Ireland. He was influenced by some excellent cherry crops on the continent and acknowledged its timber and non-timber ("as a source of sustenance for many birds and insect") benefits. There are no pure semi-natural cherry woodlands of scale in Ireland but it exists in copses and mixtures alongside other broadleaves. It requires fertile soils such as brown earths and according to Joyce needs "slightly moist soils, preferably with some lime content of a medium to high nutrient status".

Many forest owners choose cherry for its aesthetic qualities. It produces spectacular colour in spring usually lasting for no more than a couple of weeks. It blossoms in clusters with individual flowers containing five white petals and up to 20 stamens. The bark is purplish brown in colour with clearly defined horizontal pink to brown bands but eventually it turns rough and wrinkly with age.

Cherry is a ring-porous wood comprising a rich brown heartwood and pale creamy sapwood. It is easy to carve, which is why it is valued by wood turners and sculptors. It has been used in the manufacture of high-value furniture, especially since the early nineteenth century. Straight-growing, knot-free mature cherry is used for veneering and internal panelling.

Wild cherry is an attractive tree because of its profusion of white flowers in spring. Rarely planted pure, it grows well as a minority tree in mixtures.

Cherry has potential when mixed with other native species. While it is susceptible to bacterial canker, it will continue to be a minor species in woodland mixtures. The rewards are great when it reaches maturity as demand for good cherry exceeds supply. Most cherry used in Ireland is sourced in Europe or from the US where the preferred species is American black cherry.

EUROPEAN AND NATURALISED SPECIES

Many naturalised species are now regarded as Irish as they have been here for centuries. Because they were planted in brown earth soils there are no pollen records to date their introduction to Ireland. It is likely that some may have been brought here by the Normans.

Most of the species were planted widely from the 1740s onwards during the State private planting programme administered by the Dublin Society – now the RDS. In addition to the five species chosen here, forest owners have experimented with other species that grow well in Ireland, providing the right site and soils conditions are available. These include the broadleaves Norway maple, common walnut, Italian alder, hornbeam and lime. Austrian and Corsican pine also have their advocates.

The following species provide the greatest multipurpose potential in the creation of a diverse woodland: beech, sycamore, sweet chestnut, Norway spruce, European larch. They perform well either pure or in mixture, especially with native species.

Beech

Beech has a natural range over most of Europe including southern England. It failed to make it to Ireland before the landbridge, which connected both countries, was submerged by the rising sea after the last Ice Age. It is likely that beech was present here in the sixteenth century and may have been introduced earlier by the Normans but this is disputed. Like the Normans, who allegedly became "more Irish than the Irish themselves", it adapted well to Irish conditions and was adopted by Irish

BEECH WOOD

Beech produces excellent timber with a wide range of high added–value uses in solid, laminated and veneer form. Uses include furniture, joinery, flooring, domestic wood ware and turnery. It finishes well and can be shaped and bent during steam heating. It is not suitable for outdoor use.

Beech is a versatile wood for furniture-making. It can be used either in solid form or steam bent to provide flowing shapes. Alternatively, Elysia Taylor opted for beech veneers to provide a continuous flow in the rocker and side arms of her Michael Thonet's-inspired rocking chair (below) as part of her undergraduate course in Technological University Dublin after earlier exploration of wood craft in Bray Institute of Further Education.

foresters as an honorary native up until the 1960s.

Records of planting programmes in Ireland from the 1940s to the 1960s show that beech was the most widely planted broadleaf. During that period, more beech was planted than all other broadleaves – native and naturalised – including oak, ash, sweet chestnut and sycamore. Foresters even planted it in large areas such as Mullaghmeen Forest, Co. Westmeath. This 365-ha woodland is unique, not only in Ireland, but is acknowledged as the largest planted beech woodland in Western Europe. Beech was mainly planted in old estates acquired by the State as it had a record of being successfully established in previous rotations, especially during the eighteenth and nineteenth centuries. Samuel Hayes in Avondale, Co. Wicklow, was a strong beech advocate and would have influenced its species selection from the late eighteenth century as a result of his *Practical Treatise on Planting and the Management of Woods and Coppices* (1794) – reprinted in 2003 as *Practical Treatise on Trees*, with a foreword by Thomas Pakenham.

In recent decades, it has fallen out of favour and barely registers on afforestation programmes. While annual broadleaf planting increased to 35% of all afforestation from 2000 to 2022, beech accounts for less than 1%. Only 1.4% of the forests of Ireland comprise beech, many on old woodland sites. The heaviest concentration is in the midlands and east. The percentage of beech is highest in Counties Meath (11.2% of total forest area), Louth (10.0%), Kildare (3.3%), Westmeath (3.2%), Wicklow (2.8%), Laois (2.5%) and Offaly (2.3%). The emphasis today is on planting native broadleaves, and beech, although naturalised, is omitted from the State-supported native woodland schemes, but premium payments are generous for the species are generous. Thomas Pakenham has long argued for a more inclusive approach to beech conservation and afforestation in Ireland. He connected with Hayes in his foreword to *Practical Treatise on Trees* as well as commenting forcibly on the contemporary approach to beech:

> Hayes noticed the way the beech seems particularly well suited to Ireland. (By contrast, some modern ecologists advocate a kind of ethnic cleansing, in which naturalised species, such as beech are systematically hunted down and destroyed.)

Beech is an elegant tree, which retains its smooth grey bark well into old age. A few writers, such as Alice Munro and W.S. Graham, have likened its bark to "elephant skin". ("The elephant bark of those beeches / Into that lightening ..."). It has attractive pale-green leaves in spring, which are edible and can be used in salads. In this fresh state they can also be used to make syrup, which forms the base for liqueurs when combined with alcoholic drinks such as gin. The leaves darken in summer and provide a spectacular mix of colours in autumn, varying from pale yellow to ochre and reddish-brown.

It provides good yields on suitable sites but most beech

Beech, if managed properly produces excellent straight stems as in the Devil's Glen (opposite) and Brockagh, Co. Wicklow where forester Gerry Patterson maintains that, non-native tree species are now at home in Ireland.

crops will require a rotation of over 100 years before reaching maturity. Shaping or formative pruning may be necessary in the early years, as beech has a tendency to produce forked leader growth. However, a return to an increased planting density of 3,500/ha would reduce the need for this operation as a stocking of 300/ha is only required at maturity.

The biggest threat to beech is the grey squirrel, which is capable of stripping the bark of trees at all ages. Since the increase of the pine marten, grey squirrel numbers have decreased, allowing the native red squirrel to return so grey squirrel damage may be kept in check.

Planted on suitable sites, beech is well worth consideration as a woodland tree. It is a highly prized timber, capable of producing quality wood with a wide range of high added value end uses, especially in furniture.

Sycamore

Two lovers in an Irish wood at dusk
Are hiding from an old and vengeful king.

The wood is full of sycamore and elder.
And set in that nowhere which is anywhere
– Eavan Boland, from "Story"

Sycamore performs well as a single species tree (above), but is best suited to mixtures either with conifers such as larch or with other broadleaves including beech and maple, as here in Farran Forest Park, Co. Cork.

Sycamore is naturalised in Ireland and may have been introduced by the Normans, although the first recorded trees date to 1632. It is native to Central and Southern Europe with a natural range from the Pyrenees to the Alps, extending eastwards to the Black Sea. While not native to Ireland and most northern European countries, sycamore (*Acer pseudoplatanus*), has established itself so well and so vigorously that it is often thought of as a native species. Fired by her own imagination and assisted by legend and myth, Eavan Boland is entitled to locate her two lovers hiding from a "vengeful king" in a sycamore wood centuries before the species made its way to Ireland, just as she also has the freedom to set the lovers "in that nowhere which is anywhere".

Sycamore is a versatile species as a forest tree and as a wood producer. It is much more resilient than most other broadleaves against wind, salt spray and spring frosts, and grows in a wider range of sites. Although it is a survivor on a wide range of soils and sites, it excels on deep mineral calcareous soils.

Because it establishes rapidly and can be invasive, it is treated unfairly as a weed that threatens more favoured species. Leslie Andrew acknowledges its aggressive colonisation but maintains it "can support a high diversity of certain taxa, such as lichens" while "the aphids that feed on sycamore provide a resource for many animals, directly as prey and indirectly through their honeydew". It is regarded as an excellent honey source, referenced by William Wordsworth in "The Female Vagrant" as he positions the "active sire" in "His seat beneath the honeyed sycamore / When the bees hummed …".

Sycamore adapts well as a specimen tree in open spaces.

Sycamore wood is moderately dense and although it is much lighter in colour than oak or sweet chestnut, it has much of their strength properties. Its white colour, can be maintained if kiln-dried quickly after harvesting. Otherwise it turns pink-brown. It is suitable for a wide range of internal uses including flooring, furniture and turnery. The value of sycamore can be greatly enhanced if it has the unique "fiddleback" or "ripple" grain, which is highly prized for veneer in furniture and musical instruments. It is an ideal wood for use in contact with food such as bread boards, salad bowls, and butcher blocks. Early thinnings are ideal for firewood.

Sycamore leaves often contain black blotches caused by the tar-spot fungus *Rhytisma acerinum*. While it disfigures the leaf, it is not harmful to the tree. Ironically it is not as prevalent in urban areas as the fungus is poisoned by sulphur dioxide pollution, which is why sycamore and its close relatives maple and plane are often planted in built-up environments. Its biggest enemy in Ireland is the grey squirrel, which debarks the tree close to the canopy.

Sycamore is a highly adaptable species either as a woodland or shelterbelt and hedgerow tree. It also has potential in agroforestry and deserves reappraisal, especially on sites formerly regarded as the preserve of ash.

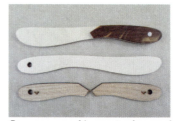

Sycamore wood is easy to clean and has tiny, almost impregnable pores, It is used for food chopping boards, and kitchen utensils such as these cleverly designed pâté and butter knives by Chaïm Factor, Hill Picket Studios. The knives are in white sycamore; the whiteness is accentuated by standing the freshly sawn planks in a vertical position before the sugars in the sap can create colour in the wood grain. The handles are presented in fume–treated and darkened Irish oak, which contrasts with the sycamore.

Sweet chestnut

Despite adapting well to Irish site and climatic requirements, sweet or Spanish chestnut is rarely planted in Ireland. It forms less than one-tenth of one percent of Irish forests, covering an area of 460ha.

It is not native to Ireland or Britain but is now regarded as a naturalised species. It was introduced to Britain by the Romans 2,000 years ago but nobody knows for sure when it was introduced to Ireland. It was probably well established by the seventeenth century as it features in *Zoilomastix* by Don Philip O'Sullivan Beare, which he completed in 1625.

Sweet chestnut is widespread throughout southern Europe but originated in what was once Asia Minor. It received its Latin name – *Castanea sativa* – from the Turkish city of Kastanaia, according to Hugh Johnson. The common name varies, depending on location, as it is known as Spanish, Portuguese and European chestnut. Sweet chestnut is probably the more ecumenical name today given its wide distribution and its ability to produce crops of edible nuts unlike the horse chestnut which is inedible, even for horses.

It grows well on a variety of sites in Ireland but it is well to avoid exposed or frosty sites. Like most species, it doesn't like extremes such as very wet and very dry soils. It needs shelter and while it won't thrive on nutrient poor soils, it will do well on moderately fertile light soils, although it may need phosphate on "nutritionally marginal sites", according to Ted Horgan and fellow authors of *A Guide to Forest Tree Species Selection and Silviculture in Ireland*.

Sweet chestnut has a long life. There are plenty of examples of trees that are well over 200 years old in Ireland. However as it grows old, it develops spiral grain and a condition known as shake, which manifests itself as cracks in the timber. The grower can opt to have fine specimen trees such as the John Wesley tree but if the objective is to produce good-quality sawable wood, then a rotation not greater than 80 years is recommended.

Sweet chestnut is an excellent woodland and parkland tree. The John Wesley Tree in Rosanna, Ashford, Co. Wicklow, is a fine example of its longevity, pictured as it appeared in 1903 (above) and in 2013 (below).

The wood is excellent and resembles oak in a number of respects; not quite as strong but easier to work. It is used for furniture and flooring and clean knot free lengths are used for veneer. It makes excellent fencing posts and providing the wood is dried properly, it will last for at lease 30 years without preservation. If the logs are ponded – immersed in water – they can last up to 40 years.

The tree has been valued on the continent as much for its edible nuts as for its wood. Our gourmets have largely ignored the fruit and while our climate is not always obliging in producing good crops, there have been excellent seed years. The nuts can be boiled or roasted and also used in savoury and sweet dishes. I remember a forest worker in Piltown, Co. Kilkenny – the late Billy Breen – used to collect the nuts in the Mountain Grove, which his mother used for a variety of dishes from stuffing to stews.

Wood of sweet chestnut has similarities with oak with a light to medium brown heartwood, darkening with age while its narrow sapwood band is creamy to light brown.

Opposite: Model School, Inchicore, Dublin: Donaghy + Dimond Architects used sweet chestnut extensively in this imaginatively designed cantilevered structure.

Norway spruce wood is white to pale yellow – known in the trade as white deal – with no discernible difference between sap and heartwood. It is straight grained with a finer texture than Sitka spruce.

In the list of Champion Trees produced by the Tree Register of Ireland it appears as a fine specimen tree not only in the aforementioned Ashford and Muckross, but right around the country. Champion trees are located in places as diverse as Scarva, Co. Down, Johnstown Castle (29m in height), Bunratty, Ballinasloe, Lough Key, the Malone Road, Belfast, Borris, Cappoquin, and Williamstown, Co. Meath.

Sweet chestnut is now an honorary native species in Britain and while it is unlikely to achieve that status here, it deserves a more prominent place in Irish forestry.

Norway spruce

The first planting of Norway spruce in Ireland "was at the beginning of the eighteenth century or even earlier", maintained HM Fitzpatrick in *The Forests of Ireland (1965)*. The most important commercial tree species in Europe, it is not native to Ireland and Britain, where it failed to colonise naturally after the last Ice Age.

Its natural range extends from the Alps across Central Europe, up to Scandinavia and across to western Russia to merge with Siberian spruce in the Ural Mountains where hybrids between the two species are found.

While Fitzpatrick believed that it might have been introduced to Ireland in the late 1600s, "reintroduced" is probably a more accurate word because there is evidence of spruce in Ireland dating to the Gortian Warm Stage (named after excavations near Gort), which is an interglacial period estimated at between 300,000 and 425,000 years ago. Frank Mitchell and Michael Ryan maintained in *Reading the Irish Landscape* that common tree and shrub species during this period such as fir, spruce, hornbeam, box and rhododendron disappeared from the Irish landscape long before the last Ice Age.

It was planted as a forest tree during the Dublin Society afforestation scheme from the mid-eighteenth century and possibly earlier. In a letter to Henry John Elwes – co-author of *The Trees of Great Britain & Ireland* – AC Forbes wrote around 1912 that he discovered a quality stand of 80-year old "common spruce, growing in Glenshiskin Wood" near Kilworth, Co. Cork.

It was widely planted in Ireland up to 1934, when it formed 34% of planting programmes, decreasing to 20% in the late 1940s and falling gradually to 2% by the 1980s. Today it forms less than 4% of the forest estate – Coillte and private.

Norway spruce has been overshadowed by its North American cousin Sitka spruce for almost 70 years, but this might be about to change. There is some evidence of renewed interest in Norway and the recent increased State grants and premiums for mixed diverse conifer forests may help a revival in planting.

The new establishment grant for Norway spruce and other diverse conifers is now €4,452/ha compared with €3,858/ha for Sitka. While forestry companies regard these grant rates as too low, the annual premium rate for Norway is attractive. It is now

€863/ha – up from €590 – compared with €746/ha for Sitka spruce – up from €510. Both carry a 20% broadleaf element but this new differential provides a farmer with a cumulative income of €17,260/ha for Norway over 20 years, which is €2,340/ha greater than Sitka spruce.

However, an attractive premium isn't the main criterion for species selection as Norway spruce is judged on its overall performance, especially against its reliable, high-yielding and market-friendly Sitka. In this regard it compares well as a forest tree.

Norway spruce has similar resistance to insect pest and disease damage as most conifers. It suffers from green spruce aphid defoliation but recovers. It is also susceptible to butt-rot which is not unusual in many conifers.

It can suffer from top-dying, which according to a 1991 Coillte information note occurs after the removal of side shelter. Heavy thinning, especially during prolonged dry conditions in spring and summer, or sudden exposure of trees after road-making or clearfelling of adjoining crop can cause top-dying.

Although not fully resistant to late spring frosts, it is much more resilient than Sitka. While frost damage was a rare occurrence in the past – once in seven years – it is now more frequent as in back to back years of 2020 and 2021. Norway may be worth exploring in low-lying frost prone sites with Sitka being planted in upland more exposed areas.

Like Sitka, Norway has a shallow rooting system and requires a sufficient supply of soil moisture. Ideally, it should be planted on well-drained moist soils, including heavy clays and gleys. Although it will perform well on moderately fertile soils including old woodland sites, Sitka is better suited to low-nutrient soils and exposed sites.

On similar sites Sitka spruce outperforms Norway in yield class (YC). The differential can be as high as YC4 (m³/ha/annum). When the age of maximum mean annual increment (MMAI) – less 30% – is factored in, YC24 Sitka is clearfelled aged 30 while YC20 Norway will be at least 40 years old. However on good mineral moist soils the age of MMAI may be less.

Norway spruce has a diverse range of end use from the tone wood of violins and cellos – sourced in the northern Italian Alps – to mass wood for buildings as high as 80m in Austria and Norway.

It is highly valued by sawmillers and woodworkers as a smooth, easily-worked timber suitable for interior joinery, unlike Sitka which is coarser. Stakewood produced from early thinning of Norway tends to be more brittle than Sitka but as both species mature to the large sawlog phase, they are in wide demand for construction. There is some evidence that sawmill rejection rates are less for Norway large sawlog than Sitka when mechanically graded.

As timber supply increases in Ireland and as we rely more on exports, Norway will find a ready market not only in the UK

Quality Norway spruce, 1912, Glensheskin, Kilworth, Co. Cork, probably planted in the early 1830s This featured in The Trees of Great Britain & Ireland *by Henry John Elwes and Agustine Henry.*

Forest educational field day and wild food foraging event in Norway spruce in Clonad Woodland, Tullamore, Co. Offaly organised by the Irish Timber Growers Association and The Irish Forest Unit Trust in 2022. During this field day, 37 different species of mushrooms – not all edible – were collected.

but also in Europe where architects, engineers and builders are familiar with the species. It is widely specified as sawn wood for timber frame houses or engineered into cross-laminated timber (CLT) and glulam as a major medium-to high-rise mass timber construction material.

In comparison to Sitka, Norway is marginally more conducive to ground vegetation cover, although both are intolerant of competition from the thicket stage when the canopy closes. Ground vegetation increases as Norway matures, especially if allowed to grow on to or exceed MMAI age.

The Woodland Trust acknowledges it as providing "a habitat for a variety of wildlife, including beetles, weevils and hoverflies" while "caterpillars of a number of moth species feed on the foliage, including the spruce carpet, cloaked pug, dwarf pug and barred red".

Norway spruce provides a rich variety of mushroom species. During a field day in 2022 in the Norway spruce dominated Clonad Forest, Co. Offaly, foragers collected 37 different species of mushrooms.

Norway spruce has potential as a major forest tree in Ireland as it is in Europe either pure or in combination with a minimum of the mandatory 25% broadleaf species and 10% open biodiverse areas.

European larch

European larch is one of three larch species in Irish forests. It was the preferred species until the 1950s when it was gradually replaced by Japanese larch, while hybrid larch, although seldom planted, is regarded as superior to both parents. European larch was introduced to England in 1620 and made its way to Ireland probably around 1740, according to H.M. Fitzpatrick in *The Forests of Ireland*. It is a native of central Europe and is widespread in the Swiss and Austrian Alps as well as the Carpathian Mountains.

Seed, if not available from Irish forests, is sourced from its natural range including the Sudeten mountains in the Czech Republic, southern Poland and the Tatra mountains in Slovakia. European larch (*Larix decidua*) comprised between 4% and 9% of afforestation programmes in the 1940s but declined rapidly as foresters looked beyond Europe for suitable species. When larch was selected, foresters invariably chose its Japanese cousin.

It is easy to understand why European larch was popular in planting programmes where species – tree and understorey – diversity is required. It is accommodating and works well in mixtures with other species. Because it is one of a few conifers that sheds all its leaves – or needles – during autumn and winter, it is aesthetically pleasing, providing good colour variation throughout the seasons as well as allowing ground vegetation to flourish. It has soil-improving qualities which result in an increase in phosphorus levels. Spruce has benefited from this "nurse effect" especially on low-nutrient old red sandstone sites where phosphorus levels are low. It grows more slowly and is more nutrient-demanding than Japanese larch. It is also vulnerable to larch canker, unlike Japanese larch, which mainly accounted for its decline. The future of all larch species is now in doubt as they are susceptible to the introduced pathogen, *Phytophthora ramorum* but European larch is more resilient and still planted. While *P. ramorum* remains a threat, larch's future is in doubt. If this can be resolved, it is worth considering in areas where species diversity is a requirement. It is a good pioneer species, establishes quickly so weed control is not a major problem. It is also a good firebreak and amenity species.

Hybrid larch *(Larix x eurolepis)* is an excellent alternative to either European or Japanese larch. It originated in Dunkeld, Scotland around 1904 as a result of accidental cross-pollination of European and Japanese larch. The artificial cross has similar botanical characteristics to its parents, but grows faster and straighter. It is an ideal choice for quality timber production but shortage of plant supply, and high production costs resulted in limited planting. Hybrid larch produces inferior, quality trees when seed is collected from hybrid larches, so seed has to be collected from an original cross to produce quality nursery seedlings. There is uncertainty about hybrid larch's resilience against *P. ramorum*. The evidence available suggests that it is hardier and more resistant to insect pest and

European larch has a higher density and stiffness than many hardwoods and most conifers, including spruce. It is more durable than Japanese larch and is the preferred species for boat-building. Internally, it is used for flooring while external applications include transmission poles, fencing, cladding and gates. The heartwood is naturally durable but sapwood needs preservatives for outdoor use.

European larch, Ballymoyer Forest, Co. Armagh, dates to the late eighteenth century.

disease damage than its parents, but the jury is out on its ability to withstand *P. ramorum*. A Forest Research species note cautiously states: "Susceptibility of hybrid larch to *P. ramorum* is uncertain at present." There is currently no research on developing resistant strains of larch in Ireland, which is a pity because it has huge potential as a forest and timber tree.

NON-EUROPEAN EXOTIC SPECIES

Ireland's exploration of exotic tree species dates to the 1830s when estate owners experimented with a number of species, mainly from western North America. Some of these early plantings still survive, such as the Sitka spruce planted in the Curraghmore Estate in Co. Waterford and Douglas fir in Co. Wicklow. Both held the record as Ireland's tallest trees up until this century when they were surpassed by trees planted as recently as the early 1900s in Avondale and Glendalough, Co. Wicklow.

Forest owners have experimented widely with diverse exotics including lodgepole pine, which was planted widely between the 1950s and 1970s, especially on raised and blanket bogs where it grew prodigiously, although the quality was uneven. Planting is no longer carried out on these sites so lodgepole pine is rarely planted today. Coast redwood and Monterey cypress grow well in Ireland while Monterey pine stands provide excellent yields and excellent commercial timber. Broadleaves such as red oak and black walnut have proven potential in the US, which could be replicated in Ireland. Irish foresters and forest owners are spoilt for choice when selecting exotic tree species but markets are likely to be poor for most of the species mentioned.

The following species provide excellent yields on suitable soils and also provide benefits as they are now planted on sites that incorporate at least 35% broadleaf species and open biodiverse areas: Sitka spruce; Douglas fir; Japanese larch; western hemlock; western red cedar.

Sitka spruce

No other tree generates more debate in Irish forestry than Sitka spruce. Foresters and forest owners regard it as the most adaptable tree species growing in Ireland, capable of producing outstanding yields, on a variety of sites including low-nutrient soils and a wide range of end products. Described as the workhorse of forestry, Sitka spruce (*Picea sitchensis*) is the most important commercial tree in Ireland and makes the greatest contribution to Ireland's forest industry. Critics point to Ireland's overreliance on Sitka, which covers 47% of the forest estate, although still only covering 5% of the land area. Where did it come from and how did it establish itself as a major species so quickly as it has been planted widely for less than a century?

It was introduced to Ireland in 1834 from British Columbia (BC), Canada. One of the first Sitka planted that year in

Female flower – ripening to a cone – with smaller male flower. Good "seed years" occur every five to seven years, usually from age 20 onwards or even earlier in fast-growing crops.

AUGUSTINE HENRY'S POSITIVE VIEWS ON EXOTIC TREE SPECIES

AUGUSTIN HENRY.
1929

Augustine Henry, forester, botanist, plant collector and joint author with John Henry Elwes of the seven-volume *The Trees of Great Britain and Ireland* (1906-13), played a major role in influencing Irish foresters and policy makers to favour Sitka spruce and other Pacific Northwest tree species in afforestation programmes.

"The fact that a species is exotic or foreign has nothing to do with the question as to whether it should be planted," he told the a government inquiry on forestry in 1907. He was conscious that Ireland would remain a net importer of timber long into the future unless Sitka was the mainstay of Irish forestry.

Henry was aware of the high performance of the species in estates throughout Ireland. When 100 tree species were trialled in Avondale between 1905 and 1908 – including nine spruce species from around the world – Sitka spruce emerged as the most versatile species for Ireland's afforestation programme in terms of climatic and soil suitability and its wide range of end uses.

Curraghmore, Co. Waterford, still survives and now measures 56m in height. Sitka's natural range stretches along a narrow coastal belt from California to Alaska through Oregon, Washington and BC, a distance of 3,600km. The tree owes its name to Sitka, the Russian capital of Alaska before the US purchased the territory in 1867. Its highest density as a forest tree is within 80km of the coast although it is found further inland where it "grows along the banks of inlets and estuaries and on steep coastal mountain sides typically to about 500m but occasionally up to 2,700m altitude" according to Ruth Tittensor in *Shades of Green: An Environmental and Cultural History of Sitka Spruce*.

With a similar latitude and maritime climate to Ireland, it's easy to see why early twentieth-century foresters concentrated Sitka seed collection in the Queen Charlotte Islands and Vancouver Island in BC and coastal north Washington, where annual rainfall can exceed 3,000mm. Later on, they would source seed in southern Washington and northern Oregon, where rainfall is lower but still comparable to Ireland. Sitka thrives in mild, moist growing conditions and is at home in Ireland where annual rainfall varies from 750mm to 1,000mm in the east and between 1,000mm and 1,400mm in the west while 2,000mm precipitation and over is not unusual in mountainous areas.

Sitka spruce is one of the easiest species to establish as it grows rapidly on moist, fairly fertile soils, but is also capable of good growth rates in exposed sites of moderate and even low fertility. Low-lying "frost hollows" are best avoided in favour of Norway spruce and Scots pine as Sitka is susceptible to late spring frosts but typical of its resilient nature it invariably recovers. It has a shallow or near-surface root system and so is regarded as vulnerable to windthrow, although research carried out by Erik Sundström demonstrated that roots can penetrate to depths of 1.5m in free-draining soils.

Since afforestation moved from peat and upland areas to lower more fertile sites, Sitka spruce, in particular, requires no fertilisers. It reaches maturity as early as 27 years but rotations exceeding 30 years are recommended to gain maximum cumulative yield and revenue. In recent years, some forest owners with Sitka forests in windfirm sites are converting their crops to continuous cover forestry (CCF) which maintains forest cover in perpetuity. State funding is available to establish CCF crops or convert existing forests to continuous cover or close to nature forestry. CCF provides continuity of income, encourages greater biodiversity and minimises soil disturbance during harvesting, especially clearfelling. While Sitka has been the preferred species of growers, foresters and sawmillers for over a half century it is not to everyone's liking. Critics, even those who acknowledge its role in commercial forestry, maintain that Irish foresters have become over-reliant on Sitka.

Foresters maintain that Sitka was the correct – often the only – species choice at a time when forestry was regarded as the land use of last resort. They also maintain that because it provides an economic return in a relatively short time, it has acted as an enabler for private forest owners to include long-rotation native broadleaves in their tree species mix. Without it many forest owners would not establish forests – native or introduced.

As better quality land is now being made available for afforestation, a more diverse tree species mix is emerging. This has been encouraged by State initiatives such as the Forest Environment Protection Scheme (FEPS) and Native Woodland Scheme (NWS). In addition, the Forest Service *Code of Best Forest Practice* along with a suite of mandatory environmental guidelines address a range of issues including water quality, fertilisation, archaeology, harvesting and biodiversity. COFORD states that non-native conifer plantations including Sitka spruce "provide suitable habitat for a wide range of native flora

Sapwood of Sitka spruce is creamy white to light yellow and merges gradually into the heartwood, which is light pinkish yellow to pale brown depending on age.

Below: Sitka spruce can be engineered into mass wood to displace concrete and steel in construction. Sitka is suitable for producing glulam as in the exposed 25-m beams in the Coillte's Beyond the Trees visitor centre, Avondale, Co. Wicklow.

Overleaf: Sessile oak seedling emerges in a Sitka spruce CCF managed research plantation in Co. Wicklow. This Teagasc-University College Dublin project – funded by the Forest Service – aims to achieve a mixed-age Sitka spruce forest over time, which will also encourage natural regeneration of native and introduced tree species.

and fauna and make a positive contribution to biodiversity conservation". These plantations "can maintain or create wildlife corridors enhancing connectivity between areas of native ecosystems". In the 2002 publication *Forestry and Bird Diversity in Ireland*, O'Halloran made a number of recommendations including avoiding large-scale blanket monoculture afforestation. Monocultures are being avoided, as the forests now emerging contain more diverse species, which has seen native species increase from 5% of annual afforestation in the last century to between 30% and 40% since 2000. While most forest owners wish to establish some Sitka spruce to ensure an economic return, they now include at least 20% native species and 15% open biodiverse areas in their afforestation plan. As new planting sites average 7ha in Ireland, this results in a maximum area of 4.5ha of commercial conifers including Sitka spruce.

This approach allows for wildlife corridors and setback from watercourses, heritage, hedgerows and other natural and man-made features even in forests where Sitka spruce is the predominant species. There are major challenges in managing Sitka spruce clearfelled sites where siltation and increased water runoff can cause pollution to streams and rivers unless proper harvesting guidelines are followed including the installation, and long-term maintenance of silt traps. After clearfell, there are now opportunities to restructure reforested sites by introducing broadleaves and setback areas.

Sitka while non-native, is not an invasive species as national inventories since 2007 demonstrate (see panel opposite) The evidence provided in successive forest inventories (National Forestry Inventory 2007-2022) clearly indicates that Sitka does not spread outside its manually planted forests.

The positive role that Sitka spruce can play in decarbonising the economy is only being fully understood in recent years. Joyce and OCarroll are emphatic about its potential as a carbon sink in *Sitka Spruce in Ireland*:

> It is a clearly established fact that the best and fastest contribution that Ireland can make to the international move to reduce the incidence of greenhouse gasses and global warming is through the planting of Sitka spruce on suitable land.

The authors concentrated on Sitka's ability to sequester carbon in the forest but its decarbonisation role in climate change mitigation extends beyond the forest in a range of products, especially construction. This is acknowledged in successive Climate Action Plans, which outline how softwood such as Sitka "can replace concrete and steel in many applications such as floors, roofs, walls and stairs due to its strength and versatility". Sitka can make a major contribution to decarbonising construction and the built environment which "are directly responsible for 37% of Ireland's emissions", according to The Irish Green Building Council (IGBC).

Sitka's diverse range of products explodes the myth that it is a low-value wood with limited uses which is still prevalent even among writers and commentators who display a deep knowledge of forestry. In an otherwise excellent history – *The Wood Age* – of humanity's relationship with wood, Rolan Ennos relegates Sitka's use to providing timber only for "sturdy pit props" in coalmines. He implies that since the mines have been closed there is no longer a market for Sitka spruce, which is patently incorrect as the species has proven itself to be a versatile timber with a huge market reach in Ireland and Britain in construction, packaging, fencing where it competes successfully with European imports. It is an ideal fibre wood for producing panel based products such as oriented strand board (OSB) and medium-density fibreboard (MDF) from pulpwood and waste wood. It is an important species in Ireland's drive to increase timber frame construction from 25% in 2022 to at least 70%.

Its main attraction for foresters is its versatility in providing

SITKA SPRUCE: NON-NATIVE BU″

An invasive species is non-native living organism that spreads naturally beyond the area where it was manually introduced and causes harm to other species and ecosystems. Typical examples in Ireland include the grey squirrel which has spread throughout Ireland and threatens the native red squirrel and some broadleaf trees. Sika deer have also spread far beyond its original confines and likewise, knotweed has established wild populations and causes damage to ecosystems and even property. So, non native species deserve to be categorised as invasive not because they are introduced but because they spread naturally beyond the area where they are manually planted and cause widespread damage.

Not all non-native plants are invasive. The introduced shrubs lilac and fuchsia and trees such as Norway and Sitka spruce, sweet chestnut, sycamore and European larch grow prolifically here but stay within the confines of their forests, woodlands or parks. I include Sitka spruce as it has been labelled invasive in *Ireland's Changing Flora: A Summary of the Results of Plant Atlas 2020*, published by the Botanical Society of Britain and Ireland (BSBI). It states: "[Sitka spruce] regenerates freely from seed, often on heathland and bog some distance away from plantations." The BSBI claim, is incorrect in labelling Sitka invasive, as is the claim by Yvonne Buckley, an ecologist and professor of zoology at Trinity College Dublin, who maintains the species "takes over heath and boglands, changing profoundly their ability to sequester carbon and provide water services". Their claims can be challenged by tracking the spread of Sitka spruce since it was introduced to Ireland in 1834, and was planted mainly in estates until the last century when it was planted widely as a forest tree. The total area of manually planted Sitka in forests now amounts to 369,900ha or 5% of the land area. There is no evidence that

ALSO A NON-INVASIVE SPECIES

the species has spread outside this area or to support the claim that Sitka "takes over heath and boglands". How can we be sure about this? We know from the results of Ireland's National Forest Inventory, carried out by the Department of Agriculture, Food and the Marine (DAFM). For example, between 2007 and 2022, the forest area in Ireland increased by 16% or 111,006ha, including a 2% increase in naturally regenerating forests amounting to (15,218ha). Natural regeneration was dominated by pioneer native broadleaf species such as birch and willow, while the per cent of Sitka spruce was so low, or non-existent, that it didn't register.

When asked in 2023, a DAFM forester said: "Limited natural regeneration of Sitka spruce occurs within planted forests but rarely occurs in adjoining bogland or heathland. Even within the forest, the Department's experience is that Sitka spruce will only regenerate naturally where favourable 'seedbed' conditions exist such as moist soils with needle litter or light moss, as can be found on sites where a conifer stand has recently been removed and when replanting has been delayed or not carried out for whatever reasons." Even in mature early 20th century forests such as Glendine, Co. Laois (pictured) and Glendalough, Co. Wicklow, natural regeneration is confined to the original planted area.

I have seen isolated Sitka spruce outside forest boundaries. These are generally the remains of once planted burned areas or abandoned exposed unsuitable deep peats, rather than the results of natural regeneration. The evidence from the team of DAFM inventory foresters who map tree species is conclusive. Sitka spruce as a forest tree and wood producer is Ireland's most important economic and carbons sequestering species. Whatever its real or perceived disadvantages are, invasiveness is not one of them.

products for the marketplace as early as year 14 when thinnings are used for panelboard products and wood energy. Within a few years medium-sized logs are provided for packaging and fencing such as pallet manufacture. The demand for packaging is huge, especially in the transport of food and beverages, IT products and engineering goods. Demand for pallets increased dramatically during the Covid-19 pandemic, especially for the transport of food, pharmaceuticals, medical devices and other goods. A major advantage of Sitka pallets is their recallability, which lowers their carbon footprint and provides savings to pallet users. Pallets are returned to be repaired for recycling several times. Because no damaging chemicals are used in pallet manufacture and restoration, they can finally be chipped for equine bedding, garden mulch or renewable energy after the end of their service life.

When Sitka spruce produces timber for construction – as early as year 20 – its full value is reached, both for the grower and the processor. Irish sawmills export most of their construction material but there is major potential to increase market share, particularly in the domestic market, as builders and architects increase timber frame construction.

Why Forests? Why Wood?

Innovative pop-up outdoor theatre using Sitka spruce pallets to create a 300-seat amphitheatre designed by Seán Harrington Architects and A2 Architects. It was constructed by young people from the Dominick Street flats in Dublin and from loyalist districts of Belfast in 2013 as part of a cross-border educational peace initiative. Recyclable timber pallets are essential in the transportation of goods and were in huge demand during the Covid-19 pandemic, especially for pharmaceuticals, food and other goods. Pallets proved to be ideal for this temporary structure because of their design. On completion of the project, the pallets were re-used a number of times before they were mulched for horticulture or converted to renewable energy.

What Trees?

An exciting new development lies in engineering Sitka to improve its strength qualities. We have seen how adaptable it has proven in manufacturing wood-based panels which bear little resemblance to its original form. Now, as demonstrated in University of Galway, it is being engineered into mass wood as a CLT product, capable of displacing concrete and steel in medium- to high-rise buildings. Sitka is also suitable for producing glulam. The exposed 25-m Sitka spruce glulam beams in Coillte's Beyond the Trees visitor centre in Avondale illustrate the enormous potential of the species while BCP Capital is planning the first CLT office building in Ireland.

Douglas fir

A native of Western North America, Douglas fir is regarded as one of the most important and valuable timber trees in world forestry. Its natural range extends from BC, Canada through coastal Washington, Oregon and California, where it is also found inland. It also grows naturally in the Rocky Mountains, South Arizona and in high elevations in Central Mexico. In its original habitat, Douglas fir can survive for over 1,000 years and has achieved a height of 120m. The tree owes its common name to the Scottish explorer and self-taught botanist, David Douglas, who sent the first seeds to the London Horticultural Society in 1827 after his expedition up the Columbia River. It owes its Latin name – *Pseudotsuga menziesii* – to Archibald Menzies, a Scottish naval doctor and botanist who discovered the species in 1791 on Vancouver Island.

It was first planted in Britain and Europe in 1828 and was

Douglas fir grows extremely well in Ireland such as in Altadaven Forest, Co. Tyrone where it is in mixture with Norway spruce.

Douglas fir mature female cones are pendulous unlike the true firs, which are erect.

DOUGLAS FIR – NEITHER TRUE FIR NOR PINE WOOD

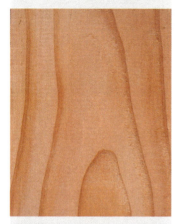

Douglas fir has an ambivalent identity as a tree and as a timber. It is known in the trade as Oregon pine, even though it is not a pine and neither is it a true fir. One of the most important conifers in the world, its timber is dense and strong and has an attractive reddish brown heartwood colour. Because it is a high-value timber, it is essential that quality is maximised for the market place.

Ideally it should be high pruned to produce knot-free logs. Douglas fir has always found a ready market for a range of high-quality end uses in Ireland, including electricity and telephone transmission poles, log buildings, joinery, garden furniture, decking, fencing stakes, cladding and veneers when pruned.

Irish foresters who wish to incorporate the species in their planting programmes would do well to visit some French Douglas fir forests. The country's 420,000ha of Douglas fir forests have a growing stock of over 100 million m³. It out-produces all other species in France – per hectare – where it has a professional association, – France Douglas – to represent Douglas fir growers, foresters and sawmillers who organise field days and support research on the species.

introduced to Ireland around 1850. The 60-m Douglas fir in Powerscourt Estate, planted in 1865, was the tallest tree in Ireland for most of the twentieth century. It is now surpassed in height by three Douglas fir trees in Avondale and Ireland's tallest tree, a Sitka spruce in Glendalough, all planted in the beginning of the last century.

The total area of forest under Douglas fir was 9,300ha in 2022 compared with 10,380ha in the 2017 National Forestry Inventory. This is a surprisingly low and a disappointingly declining area, especially as excellent stands have been established over the years throughout Ireland especially in Counties Wicklow, Laois, Louth and in the old red sandstone (ORS) valleys of the Suir, Nore and Blackwater rivers. There were also some fine stands in Northern Ireland – the first State afforestation in the North was a 4.5ha plantation of Douglas fir in Caman Wood in 1912. C. S. Kilpatrick recorded that this plantation "matured and was felled between 1962 and 1969" in Northern Ireland Forest Service – A History. About 3% of all afforestation comprised Douglas fir in the 1920s and 1930s but it fell below 1% thereafter as Sitka spruce and lodgepole pine gained prominence. Douglas fir was widely planted in Europe where it now covers an area of over 767,600ha comprising France (420,000ha), Germany (217,600), UK (45,000ha), Spain (25,000ha), Belgium (23,000ha), Netherlands (19,000ha) and Slovenia (18,000ha).

It grows extremely well on suitable sites and is capable of producing high-quality timber in 40-year rotations. Douglas fir performs best on well-drained, moderately rich soils. It requires shelter and grows best in warmer and drier parts of the country as recommended by Horgan and co-authors in A Guide to Forest Tree Species Selection and Silviculture in Ireland. Good vegetation indicators are grass/herb, and fern/grass. However the authors urge caution on some bracken sites.

Growers should avoid planting Douglas fir on exposed sites. It is suited to sheltered, middle-valley slopes and the Forest Service recommends that it should be limited to elevations below 200m, although on some sites in Co. Wicklow it performs well up to 300m altitude. While the species tolerates low winter temperatures, it can be damaged by late spring frost.

Douglas fir also demonstrates its ability as a nurse species for Sitka spruce in less fertile old red sandstone upland areas. A nurse enables an adjacent species to grow satisfactorily on what would normally be an unsuitable site. This function can be achieved by protecting the other species from frost and exposure, or in the case of Douglas fir, passing on much-needed nitrogen to the nursed species, especially on low-nutrient soils. Research experiments show an increase in yield class of almost 70% in Sitka spruce planted close to Douglas fir in nitrogen-poor peaty mineral soils derived from ORS. However, it is unlikely to be called upon to fulfil this function as more productive sites are now available.

It requires little site preparation and vegetation competition in the early years is not a major problem due to its rapid early growth. It is relatively free of diseases and pests. Fertiliser application is usually not necessary, although it is recommended that young plantations are monitored regularly, especially for phosphate levels. Many foresters and growers have shied away from planting Douglas fir because of its susceptibility to deer damage. This problem has not been resolved in areas of high Sika deer populations.

In the past, a lack of suitable sites restricted Douglas fir planting programmes. Better-quality sites are now available so there are greater opportunities for species diversity. Against this background Douglas fir is worth consideration by growers who wish to establish species that will produce high-quality durable timber with a wide range of added-value uses.

Japanese larch

Introduced to Ireland in 1861 by John Gould Veitch according to HM Fitzpatrick, the natural range of Japanese larch is in Honshu Island, Japan, but it grows naturally throughout the country. That Veitch introduced it himself to Ireland may not be correct but he had introduced it to Britain by 1760 so it is likely to have arrived here shortly after as estate owners and managers were keen to feature non-indigenous trees in parks and gardens. it has a number of qualities that make it an attractive option both as a decorative and commercial tree.

Douglas fir was used for the Viewing Tower in Avondale because its high resin content means it has great ability for self-protection, so it does not need to be treated with chemicals in above ground use.

Opposite: Japanese larch is one of just a few conifer species that sheds its needles during the dormant season.

Japanese larch (*Larix kaempferi*), in common with European and other larches, is deciduous, unlike most conifers. It
has most of the characteristics of European larch, with its similar bark and soft, feathery bright-green needled tufts in spring.
It has reddish brown shoots in autumn while European has less
distinctive straw-coloured shoots. Because it allows sufficient
light through during the dormant season and doesn't have the
heavy canopy of other conifers during the growing season, it
is conducive to good ground cover of grasses and plants. It is
therefore an amenable and pleasant tree, with environment-en-
hancing qualities. It also can be planted in mixtures with other
species, where it acts as a nurse.

It vied with European larch in afforestation programmes in
Ireland until the late 1950s but greatly exceeded it thereafter.
Today, 22,140ha of Japanese larch forests exist compared with
2,080ha of European larch. There are a number of reasons why
it surpassed European larch in popularity: it produces higher
yields, is resistant to larch canker and as Fitzpatrick pointed
out "It grows in soils too poor for European larch and stands
exposure well". However, it may be it more vulnerable to *Phytophthora ramorum* and, as a result, it no longer features as a
State grant-approved species.

*Japanese larch, Cruagh Forest, Co.
Dublin.*

It can grow too vigorously leading to coarseness and poor form, but generally has sufficient numbers of good-quality trees to form a commercial crop. Stake producers and boat-builders prefer European larch because it is more durable and generally straighter. Still it is sufficiently durable to be used for external cladding, garden furniture and transmission poles.

Western red cedar

Western red cedar's natural range is similar to Sitka spruce, as it extends close to the coast region from Alaska to northern California. It is an important but not dominant forest tree in British Columbia and the US states of Washington and Oregon, where it grows in association with Sitka and other western North American natives such as Douglas fir, as captured in David Guterson's novel *Snow Falling on Cedars* which is set in Puget Sound. Irish foresters experimented with western red cedar in the early years of State forestry and there are excellent stands of the species dotted around the country.

Like other western North American commercial species, western red cedar (*Thuja plicata*) gained popularity as a specimen tree when introduced in 1853 by William Lobb, according to HM Fitzpatrick. He compiled a list of mainly private estates where large individual trees were – and some still are – growing prolifically but also cited excellent cedar forests in Ballykileavan, Stradbally, Lackendarragh and the Co. Wicklow forests of Avondale and Enniskerry.

Western red cedar has a whitish sapwood and dark chocolate-brown heartwood. It is naturally durable so it is used outdoors as in the Davagh Dark Sky Observatory located in Davagh Forest, Co. Tyrone, designed by ARCEN Architects.

The species responds well to Irish climatic conditions which are not dissimilar to its natural range particulary in Washington State and British Columbia where there is copious rainfall, a mild climate and long growing seasons. It can withstand the occasional low winter temperatures and the more frequent late spring frosts we get in Ireland. Cedar's failure to make it as a popular forest tree was mainly due to a thuja fungus – *Keithia thujina* – during nursery seedling production, according to Fitzpatrick, while Horgan and fellow authors report that this disease can also attack forest trees. However, they say it is largely disease-free and is worth planting, because:

> It is clearly an accommodating species, succeeding on a wide range of soils, including some nutritionally marginal, Old Red Sandstone podzols, flushed or reclaimed peats and alkaline soils.

It has potential as a nurse species in protecting neighbouring trees. It is ideal in mixtures where its slender form and soft foliage doesn't damage neighbouring trees – broadleaves or conifers.

Cedar is an attractive timber with a whitish-cream sapwood and dark chocolate-brown heartwood which ages to reddish brown and weathers eventually to a silver-grey colour. It has a straight even-grain and is very durable. The wood is acidic and corrodes most metals when damp, so fittings such as stainless steel, silicon, bronze or copper should be used. Because it is naturally durable, it is ideal for external cladding, posts and poles, shingles, boat-building and decking. For that reason, and its adaptability as a forest tree, it is worth considering as a minor coniferous species.

Seven-year-old naturally regenerated western red cedar, in the Devil's Glen Forest, Co. Wicklow. This species along with nearby western hemlock would eventually outgrow neighbouring birch, Japanese larch and beech unless the forest is managed according to the principles of continuous cover forestry (CCF). This silvicultural system is now being practised here by Coillte, to ensure equilibrium between native, naturalised and exotic species.

The cedar used in this extension was chosen for its durability and also to act as a counterpoint to the timeless rendered stonework and simple rustic concrete-work of the original house and outbuildings. The façade and roof cladding are untreated and are left to weather.

Western hemlock

There are sufficient stands of western hemlock in Ireland to suggest that it adapts well to our wet maritime climate. Like many of our Pacific North West exotics, western hemlock (*Tsuga heterophylla*) follows a similar natural range from southern Alaska to northern California, but it grows well on both sides of the Rocky mountains. Unlike species such as Douglas fir, it rarely grows in large-scale pure stands but responds well in mixture where it can act as a nurse species.

An accommodating species, it withstands damage from surrounding trees. But neither does it cause damage to nearby trees because of its light branches and soft fine needles. As a result, it grows well in mixtures with conifers and broadleaves but can eventually outgrow all neighbours unless carefully managed. It grows well in shade and is especially vigorous when the canopy opens to light. It has a bushy form when

The wood of western hemlock is even-grained and pale yellowish brown with little differentiation between sapwood and heartwood.

grown on exposed sites but in sheltered sites, especially in mixtures, it is the most graceful and slender of trees. In pure stands, it is intolerant of ground vegetation but as it is confined to small stands it has a gothic monumentality that is visually impressive, as seen in sites such as the Devil's Glen Forest in Co. Wicklow.

Many of the western hemlock stands have been removed from Irish forests and only 340ha of pure stands now exist. It is also found in mixtures but needs careful management as it can outcompete other species, especially broadleaves.

Its wood is non-resinous, even-grained and pale yellowish brown§ with little differentiation between sapwood and heartwood. It is non-durable and resistant to preservatives, so it is limited to internal construction and joinery. Straight and even-grained, it is regarded as superior to Sitka spruce in terms of strength and stiffness. It has a finish similar to Norway spruce.

Western hemlock is a prolific natural regenerator. Unlike most conifers grown in Ireland, it is a shade bearer and will regenerate rapidly as soon as there is the slightest opening in the forest canopy.

However, it lacks the ease of establishment and marketability of these species but it is worth planting in small stands, especially in mixture.

It was always regarded as a minor species by Irish foresters even though it grows well in sheltered sites and reasonably fertile mineral sites, but will also perform well on infertile mineral soils where it can take time to establish sufficiently. Western hemlock is susceptible to frost. It can suffer damage from Fomes butt-rot but is largely disease-free.

While Ireland has a limited menu of native species, the list of 16 native, naturalised and exotic trees featured here is a guide to what grows well in our mild, moist maritime climate. We have a multitude of choices when selecting tree species. In addition to this list, foresters and forest owners, especially farmers, should also explore other species, particularly the natives, which form an integral part of their landscape and ecosystem. Native species such as, holly, hazel, crab apple, willow, blackthorn, hawthorn, aspen, whitebeam and arbutus (although its nativeness is now contested) may not be forest trees but they should be planted in areas of biodiversity or setbacks or they can be integrated in woodlands and forests, especially where continuous cover forestry is practised.

The expectation in the Forestry Programme (2023-27) is for a national afforestation target of 50% broadleaves – mainly native – may be questionable, but if it is to be achieved, it needs to be based on science and research. It presents a major challenge but also an exciting opportunity in the long term to create productive forests in harmony with a vibrant native woodland resource. However, all the scientific evidence is clear that during the period up to mid-century, commercial coniferous species will have a major role to play in climate change mitigation. Achieving multifunctional forests will require an acceptance of all tree species that flourish here but may originate elsewhere. This requires an acceptance of species diversity including the need to shake off a "native-only" approach, like we have in agriculture, horticulture, sport and the arts, especially music and literature. Just as a multicultural environment encourages diverse voices to combine and create a distinctive Irish tradition in the arts, a multipurpose forest resource can create a new forest ecosystem that is also originally and distinctively Irish.

CHAPTER NINE

Our Future Forests

A forest is language; accumulated years.
Murray Bail, *Eucalyptus,*1998

Wood is an extraordinary renewable resource and taking it from well-managed sources, benefits forests and the planet. But on their own, natural forests cannot provide all the wood that we need. So, we also have to farm trees, just like we do other crops and create a new generation of plantations; plantations that allow wildlife to pass through natural forest corridors, that benefit local communities and economies and that are planted on existing cleared land, so they don't replace natural forests.
David Attenborough, *Our Planet: How to Save our Forests*, 2019

The past century is unique in the history of Irish forestry. It is the only period, since the fourth millennium BC, that forest cover has increased exponentially. The land area under forest cover, has now reached 11.5% after falling to around 1% when the State was founded in 1921. This ten-fold expansion was driven by a need for self-sufficiency in timber after four centuries of almost total dependency on imports. It was achieved by a State land acquisition and afforestation programme that lasted for most of the last century until it was replaced by the private sector – mainly farmers. Farmers in particular were incentivised to convert some of their land to forests for a variety of reasons. It would lead to a more balanced State-private forest estate and encourage afforestation on better quality – albeit marginal – agricultural land. However, the incentivisation of farm forestry was due to agricultural policy changes within what was then the European Economic Community (EEC).

Conversion of agricultural land to forestry was one policy change by the EEC to offset the massive agricultural surpluses of butter, milk and beef that had built up during the 1980s. Thus, forestry policy in Ireland has been determined by politics – domestic and European. This has influenced forest ownership and structure as well as the scale and forest types that have emerged, especially since Ireland joined the EEC in 1973.

Productive forests have created a resource that is the basis for the development of an international-scale forestry and forest products sector due to the establishment of the State company Coillte and the emergence of a competitive timber processing industry. It has also created a more balanced private-State forestry sector since farmers in particular took over the mantle of afforestation from the 1990s while the State's open forest policy initiated in the 1970s and continued by Coillte has provided the greatest recreation resource available in Ireland.

Ireland's State-owned forests and woodlands provide a sense of ownership among the public which is different from any other land use. As a result, this transformation has created greater expectations from our forests and our future forests. These include the role forests can play in biodiversity enhancement and in climate change mitigation as identified in Ireland's Climate Action Plan (ICAP) which aims to achieve net-zero greenhouse gas (GHG) emissions "no later than 2050". ICAP identifies afforestation "as one of the largest land-based, long-term climate change mitigation measures available to Ireland". It also acknowledges the role of timber production including the "continued increased use of harvested wood products in the built environment" to displace concrete and steel, while it also supports sustainable production of wood biomass for energy in "contributing to the reduction of fossil fuels".

How forestry is equipped to meet "the climate and biodiverse crises" is outlined in Ireland's Forest Strategy 2023-2030. According to the strategy, this will require future forests to deliver a diverse range of wood and non-wood products and services, encapsulating:

> multiple benefits for the environment, economy and society; the emerging circular and bioeconomy; wood product innovation and greater use of timber in houses and buildings; rural development and job creation …

The strategy acknowledges the scale of this challenge which it says, "will require a whole of society and a whole of government response if we are to succeed". Within the "whole of society", forest owners and potential owners will ultimately ensure the success of the strategy but as in the past, their decisions and outcomes will be heavily influenced by political decision both in Ireland and Europe. There is an extra factor this time around which was absent, especially during the largely State afforestation of the last century. While forest owners and foresters previously managed their forests to achieve largely commercial objectives, they no longer have carte blanche in dictating Ireland's arboreal future.

Now, other voices also influence the course of Irish forestry. These include tree and woodland enthusiasts, community organisations and the general public who use forests as places of recreation and who expect to be consulted on forest planning.

SURVEYS ON PUBLIC ATTITUDES

The following two surveys carried out by the Department of Agriculture, Food and the Marine are important in assessing the public mood on forestry:
a) Public attitudes to forestry – conducted by Behaviour & Attitudes (B&A); and
b) M-CO survey to help create a "Shared Vision" for Irish forestry.

The B&A survey found that "Three out of four people were in favour of planting more forests in their county, as well as more trees in urban areas". The survey identified the most important benefits of forests as: climate change (42% of respondents); enhanced air quality (15%); contribution to wildlife and biodiversity (14%); and mental health improvement (14%). The survey found that 88% thought forests and woodlands benefited local communities. Regarding forest structure, especially tree species mix, the results showed: Given a choice between planting broadleaf or deciduous trees, or alternatively, coniferous or evergreen, there tended to be comparatively even preference.

A slightly greater number, who expressed a preference for one type or other species, nominated broadleaves. Nonetheless, the vast majority felt that a mix between conifer and deciduous was preferable. This result "slightly surprised the researchers" who stated: "We perhaps naively assumed that there might be a greater preference for broadleaf [trees and forests] but ultimately this was not borne out by the research, with most content to say that they like either type of tree, not preferring one ahead of the other." Regarding the local contribution of forestry, only 3% "of respondents suggested forests and woodlands do not provide benefits to their local community".

CONSENSUS THROUGH CONSULTATION

The survey, conducted by M-CO, was carried out to inform DAFM and stakeholders in their quest to create "A Shared Vision for Forestry and a New Forest Strategy". In reply to the question: "Do we really need more forests at all? 97% of respondents said they were in favour of more forests in Ireland. In a follow-up response to the statement "Ireland needs more forests, because ...", the survey analysis identified climate change (carbon sequestration/sinks and climate adaptation such as flood prevention and soil integrity) and nature (mainly as biodiversity, improved habitats and ecosystems) as the dominant themes.

These were supported by wider environmental benefits (or ecosystem services) such as improved air quality. The survey also pointed out that "a large number of responses discussed the desire for increased native and broadleaf species with a significant discussion around the tensions that can exist about non-native species". The "right reasons" for forest creation included nature (98%), climate change (98%), people (94%), wood (80%), aesthetics (78%) and rural development (78%).

The majority of respondents who access forests for recreation would like to see more native species, whereas landowners with forests and those employed in the forestry sector favoured greater species diversity.

In the "Right Reasons – Forests for Wood" section, 69% preferred more timber usage in Irish homes, with only 5% disagreeing. However, 50% disagreed that Ireland needs more non-native and conifer trees, which are essential for decarbonising construction.

Trying to achieve consensus on the direction forestry should take among such a broad church is a bigger challenge in Ireland than in most other European countries, where there is little room for forest expansion and where forests have been topographically fixed for generations.

What the public and specialist interest groups want from forestry and what forest owners, foresters and downstream timber industries can sustainably deliver is the key to developing a balanced forestry programme in Ireland. Trying to achieve consensus between all stakeholders is a major challenge facing Irish forestry as there are widely conflicting views on what forests should – or should not – deliver.

Listening to stakeholders' views at local and national level is now an intrinsic element in developing forest policy. Forest owners and foresters who sustainably manage forests to achieve independently verified forest certification are obliged to listen to the views of stakeholders (see Chapter 4). The State also has a duty to listen, especially when developing policy and providing supports in shaping our future forest resource. The State hasn't been good at this, but in recent years it has been listening to the diverse – and contrasting – views of all stakeholders and as a result, there is now a greater understanding of what people expect from our forests. From this consultation, Ireland's Forest Strategy 2023-30, emerged in 2023. This involved consultation with forestry stakeholders in Project Woodland, a two-year series of stakeholder meetings and two independent surveys carried out in 2022 for the Department of Agriculture, Food and the Marine (DAFM). These provided insights to people's views on forestry and the kind of forests they favour, including the services and goods that they should provide (see panel).

Regarding species preference, the results surprised the pollsters, as the vast majority of respondents preferred a conifer-broadleaf mix. It's possible that there is now greater awareness of species diversity due to high annual recreational use of Irish forests by the public – 38 million visits, north and south. This may result in a greater awareness of species mix, as most of these visits are to predominantly coniferous forests.

There was ambiguity in replies, especially to the M-CO poll. For example, 69% preferred more timber usage in Irish homes, yet only 27% agreed that "conifers should be planted to achieve the goal of more timber usage in construction". This view is similar to a poll carried out by RedC for Coillte which found that there is broad agreement on the importance of forests to produce "more wood products that can substitute for carbon intensive materials … in mitigating climate change". Yet, there was also a strong preference to plant broadleaves and native

Ireland's recreation forests and woodlands, which are mainly managed by Coillte, provide a sense of ownership among the public which is different to any other land use.

species rather than conifers, which are the only species that can displace fossil-based material in construction, the greatest global emitter of GHG. These conflicting preferences suggests that there will be tensions on species choice, which is acknowledged in Ireland's forest strategy. It also suggests that there is a greater need for a public awareness campaign on timber usage and suitability for both conifers and broadleaves in achieving a balanced forestry programme.

While views diverge considerably on the future course of Irish forestry, there are areas of agreement. The mid-century objective of achieving carbon neutrality has concentrated minds on achieving 18% forest cover in combating climate change. It is no surprise therefore, that there is an overwhelming majority in favour of more forests as expressed in the findings of the 2022 "Deliberative Dialogue Forum" conducted by DAFM. When the 99 members of the forum, were asked to vote on the level of increased forest cover required, they responded as follows:

> 36% ... agreed that the aim should be to achieve 18% forest cover by 2050 or sooner, while a further 52% agreed that the aim should be to achieve an even greater expansion of forest cover to help combat climate change. Taken together, 88% of participants endorsed a forest cover target of 18% or more by 2050.

In contrast to public opinion polls on forestry, there are few surveys on what forest owners – and potential forest owners – want from their forests. A survey carried out by the Royal Dublin Society (RDS) on forest owners, shows a wide range of objectives (Figure 9.1). In this study, 26 objectives were listed and rated on a multiple points system totalling 632. The top five preferences were income generation (18%), diversification (15%), environment (9%), habitat for wildlife (9%) and social amenity (7%).

The results of the RDS survey are important because the poll was carried out among entrants to the Society's forestry award who manage their forests to fulfil a wide range of sustainable objectives. While revenue generation and diversification [of income] are prioritised, the poll – like the awards – reflects a diversity of objectives among entrants. The awards have evolved over the years to reflect this diversity, which is matched by the award categories: farm forestry, native woodland establishment and conservation, production forestry and community woodland projects.

The RDS survey featured the views of farmers and non-farmer forest owners, but very few surveys have been aimed directly at farmers. Taking their views for granted has proved to be a major mistake. Farmers, for example, who were the backbone of a major afforestation drive from the early 1990s

Figure 9.1: *Forest/woodland management objectives of entrants to RDS-Forest Service Forestry Awards.*

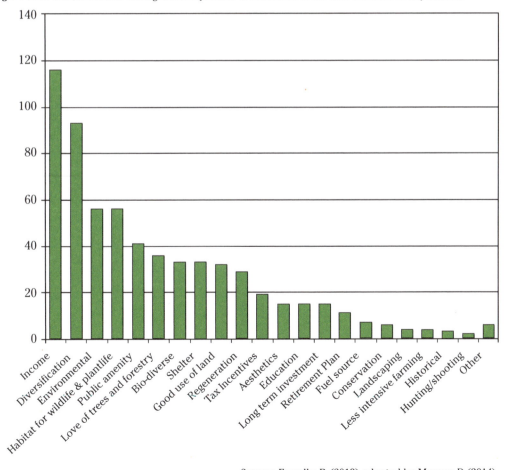

Source: Farrelly, P. (2013) adapted by Magner, D (2014).

until 2012 are now more cautious when choosing forestry as a land use option. As a result, the afforestation programme has dramatically declined, and farmers' share of afforestation has also fallen.

Why have farmers left the afforestation programme and more importantly what needs to be done to attract them back? Despite a dearth of surveys on their views, there is sufficient information available to indicate what they want from their forests. An IFA survey published in 2013, featured the views solely of farmers. Importantly, half of the farmers surveyed owned their own forests. The IFA survey was carried out "to determine the reasons farmers decided to plant, as well as to identify the barriers to future planting". Michael Fleming – a forest owner, farmer and forestry contractor and then chairperson of the IFA farm forestry committee – summarised the reasons why farmers plant and don't plant:

> The primary reason farmers plant is the forest premium (45%). The other main reasons are the good return from forestry on marginal land (33%) and investing in their retirement (16%). Of the farmers that had planted in the past, 47% said they would not plant again. The 8% cut to the forest premium in 2009 has deterred many farmers from planting; they no longer see the forest premiums as guaranteed and this has damaged confidence in the afforestation programme. The other main reasons are the lack of available land suitable for forestry (35%) and that land is needed for agriculture (15%). For farmers that decided not to plant, the lack of available land was the main reason for their decision (62%). Insufficiently attractive forest premiums (13%) and the permanence of the decision to plant (11%) are also key factors deterring farmers from forestry. The replanting obligation and the long-term nature of the investment, combined with the lack of confidence in the continuation of the forest premiums is impacting on afforestation rates.

The IFA survey highlighted areas of concern when it was being carried out in 2012 as private afforestation had fallen to 6,652ha from 15,054ha 10 years previously when an average of 12,000ha of planting was carried out by farmers. At that stage, farmers were still planting most of the afforestation programme but this has now fallen considerably, with institutional and other private investors filling the vacuum.

So who will create and manage the 450,000ha of new forests required to meet Ireland's Climate Action Plan target? I have deliberately afforded considerable space to determine farmers' views on forestry because without farmer involvement, Ireland will not achieve a viable forestry programme. Forestry and agriculture are interlinked in achieving climate change targets but presenting forestry as the only alternative to conventional agriculture at a time when farming is constantly being reminded that it accounts for 33% of Ireland's total GHG emissions can be counterproductive. While this sword of Damocles hangs over farmers' heads, all alternative land uses, including forestry, will be viewed as a threat to their livelihoods, especially among dairy and meat producers who will argue that in a global context they are grass-based sustainable food producers. Teagasc research shows that Irish dairy farming has the joint lowest carbon footprint in the EU alongside Austria, while beef production has the fifth-lowest. Even before considering forestry, farmers have the capability to further reduce GHG within conventional agriculture. Initiatives include the introduction of new technologies in GHG reduction, improving animal genetics to lengthen the grazing season, anaerobic digestion to create biogas/biomethane, switching to "protected urea" fertiliser and planting clover to reduce chemical nitrogen fertiliser. Taking the land out of agriculture is a controversial option, which is why farmers are sceptical about rewetting, which depends on State subvention in perpetuity and forestry, which requires partial State aid. In his study, *The Economics of Afforestation and Management in Ireland: Future Prospects and Plans*, Prof. Cathal O'Donoghue, University of Galway believes that a large-scale shift to forestry represents the best option:

> Without a major afforestation strategy, it would be impossible to achieve carbon neutrality objectives using rewetting or agriculture alone unless there was a major reduction in animal numbers with consequential economic impacts.

This proposal will be more acceptable to farmers if they see forestry as a viable economic and ecological proposition; economic for commercial forests and ecological for long-term broadleaf and native tree establishment. Notwithstanding the barriers to farm forestry, there is also a willingness by farmers to get involved in tree planting according to the study *Irish Dairy and Drystock Farmers' Attitudes and Perceptions to Planting Trees*, conducted by UCD and Teagasc. The authors – Rachel Irwin, Áine Ní Dhubháin and Ian Short – found that "82% of participants intended to plant trees on their land in the next five years".

While there is an overwhelming desire to increase forest cover, there is also agreement on the role forests play in recreation. "Forests are seen as hugely important in terms of providing spaces for both outdoor physical activity and also for mental health and wellbeing," according to Coillte's RedC poll. Regarding the contribution of forestry to local communities, respondents to the B&A survey were extremely positive with "just 3%" [suggesting] that forests and woodlands do not provide benefits to their local community.

While there is unanimity on increased forest cover and the role forests play in recreation and community wellbeing, the real debate begins when trying to get consensus on future forest structure, such as tree species composition. While views vary on species selection. there is a clear preference for greater diversity. Future forests need to be more diverse but also sufficiently productive to maximise carbon sequestration in the forest and to satisfy the need for more wood usage in displacing non-renewable carbon intense materials in construction and energy. These have been outlined by Prof. John Fitzgerald, as well as other issues in an expanding forest estate:

> [Forestry] must provide a clear and certain economic return, otherwise it won't be sustainable. Forests need active management. Failure to manage forests has been a major factor in disastrous forest fires. New forests also need to be robust in the face of global warming – the species chosen should be able to thrive in a warmer Ireland. New forests should not be over dependent on a single species. We have seen how elm and ash have been severely affected by disease. Finally, the expansion of forestry should be planned so that it enhances biodiversity rather than reduces it.

This echoes Attenborough's contention that "we need wood" and that we "have to farm trees, just like we do other crops and create a new generation of plantations". But Attenborough maintains we have to do this wisely so that we create "plantations that allow wildlife to pass through natural forest corridors ...". To achieve this aim, Ireland will need to increase and enhance its depleted natural forests, but there are significant opportunities to develop and enhance other natural corridors such as hedgerows and open biodiverse areas alongside our considerable network of watercourses.

If Ireland is to achieve greater biodiversity – tree, plant and wildlife – within its forest estate, it needs to place a value on this. If the goal of the current government strategy is to achieve at least 50% native tree cover nationally, and at most 65% productive forest cover at local level, then a value has to be placed on enhancing biodiversity by setting aside more than a third of their forests. Likewise, as carbon neutrality is an imperative by 2050, a value needs to be placed on carbon. In this regard the Society of Irish Foresters (SIF) proposes a forestry scheme:

> that will leverage the carbon value sequestered by forests and use this to reward and incentivise farmers and landowners to undertake afforestation. Up until now the State has quietly recorded the carbon sequestered through afforestation in its national carbon accounting while giving no reward to forest owners.

SIF advocates a carbon based afforestation scheme that acknowledges not just the environmental benefits of carbon forestry but also the monetary value of carbon sequestered by forests. This would provide an additional income to forest owners that will continue years after the premium payments have ceased. The SIF proposal has merit not only because it provides a sustained income for "public good" forestry, but it also renders the replanting obligation argument redundant.

Three key questions remain in developing a forest resource that helps Ireland achieve carbon neutrality by mid-century.

The first two relate to forest location and what kind of forests are best suited to Ireland, while the final question is how can Ireland move from a business-as-usual approach that has seen the collapse of the forestry programme, to a dynamic new model that will allow forestry to claim centre stage in climate change mitigation and rural development.

FUTURE FOREST LOCATION

The word 'forest' is derived from the Latin foris *'out of doors':*
the land that lay beyond those enclosed for agriculture or parkland.
Michael Allaby, *The Concise Oxford Dictionary of Botany,* 1992

Travel through the uplands and mountains of much of central and northern Europe and the meaning of forest or *foris* is obvious. Forests are associated with poorer land and generally comprise conifers on the exposed uplands as opposed to woods or woodlands, which mainly carry broadleaves in the lowlands and often lie between farming land and the coniferous forest. Conifers grow on the poorest sites and the highest altitudes, eventually ceasing to grow at the "tree line" where soils are too shallow and temperatures too low to support forests. In some European mountain ranges such as the Swiss Alps, Tatra and Sierra Nevada, altitudes can range from 1,600m to 2,400m while the tree line in the Andes, South America, is over 5,000m. Early plant collectors who went to the USA and Canada would have been impressed with conifers growing to altitudes of 3,550m in the Rocky Mountains. The seeds collected for Irish planting programmes were sourced at lower altitudes in western North America but even here many of the species we grow in Ireland performed well on the poorest of soils.

The definition of forest as land on the outside – that is outside the managed and tamed land – is clear in western North America and most European landscapes. The distinction between forest, woodland and agricultural land is less clear in Ireland mainly because the country was denuded of forests and woodlands by the beginning of the last century.

The large tracts of pine forests that covered much of the landscape began to die off due to a combination of man-made clearances and fire in combination with the spread of bogland 7,000 years ago. The demise of the native broadleaved forests was slower but most of these were wiped out by the early nineteenth century. The pine forests that grew at high altitudes are long gone, so in Ireland we have little concept of the natural forest or what constitutes a tree line.

However, Ireland's pioneering foresters well understood the meaning of the word "forest". The seed they collected from pine, spruce and other coniferous seed stands in western North America was used for afforestation programmes on land well outside traditional farming land or enclosed land. While the term "tree line" may not have concerned them, they established some plantations close to what might have been a tree line for Ireland. For example, high quality forests were established in the Wicklow Mountains at altitudes up to 550m.

To begin to answer the question of where future forests should be located, an evaluation of the early and more recent history of Irish forestry – successes and failures – is essential. Forest establishment and management techniques achieved the aim of large-scale job creation in rural areas as well as highly productive forests, many of them in inhospitable upland sites. Conversely, large-scale social forestry in western blanket peats, designed to create employment when emigration was rife, proved to be economically and ecologically unsustainable.

While good forest sites, including former woodland estates, rushy sites and fertile uplands were planted, forestry was invariably forced into the poorest of sites including very high altitude exposed sites and deep peats. Caught between a rock and a soft place, forestry was known in Ireland as "the land use of last resort". Forest development for much of the last century was influenced by political decisions. The emphasis was on social forestry in order to provide jobs, often to subsistence farmers who lived close to newly established forests. In addition, an unwritten but implicit land use policy of sorts ensured that forestry should never compete for agricultural land, even marginal land that was unsustainable for farming. Despite being located on largely inhospitable landscapes, magnificent forests were created.

It is against this background that twenty first-century forestry has emerged. Afforestation is subject to a wide range of soil, site and environmental conditions, which have impacted on where and how forests are established. Afforestation no longer takes place on exposed upland sites and blanket bogs. If planting is carried out on peaty soils, peat cannot exceed 30cm in depth where commercial forests are established. Afforestation is prohibited on virtually all unenclosed upland, even on productive sites where environmental restrictions don't

The virtual banning of afforestation on unenclosed land in Ireland has destroyed any chance of recreating the primeval forests of Scots pine and oak on the uplands and lower slopes, such as the Coronation Forest which has potential to be restored as a major native woodland.

apply. Afforestation sites have to satisfy a wide range of environmental conditions. For example, new planting is banned within the top eight pearl mussel catchments, while setback distances apply on all watercourses. Protecting bird habitats is a precondition on all afforestation projects. These include an afforestation ban on 167,000ha of hen harrier designated areas, while planting within or close to breeding wader hotspots will require ornithological reports.

Forests are now planted in lowland, more fertile sites, while environmental conditions attached to afforestation projects, though necessary, have slowed down afforestation. However, a number of blanket diktats by the Forest Service on where afforestation cannot take place is designed to seriously inhibit forest development in Ireland. Land availability was identified by farmers as a major factor that inhibits afforestation, according to Michael Fleming:

> The strong association between forestry and marginal land is clearly evident from the [2013 IFA] survey. If the afforestation targets are to be achieved, we must open up the land bank of good marginal land that has a proven capacity to grow quality commercial timber. The Forest Service decision to introduce a maximum threshold of 20% on unenclosed land per application, an arbitrary rule that has no basis in science has negatively impacted on the afforestation rates. It has been Forest Service policy to progressively restrict unenclosed land even prior to the introduction of the 20% rule.

Fleming acknowledged that much unenclosed land is unsuitable for forestry, but some contains "good marginal land with a proven track record of growing commercial timber [to] maximise the productive capacity of Irish farms". His views were supported in a 2015 report by Dr. Niall Farrelly and Dr. Gerhardt Gallagher. They maintained that "it may be necessary to consider all sources of land [for afforestation], including a significant area currently under-utilised (unenclosed land) ..." which they calculated as "178,000ha" or 2.5% of the land area.

Ray Ó Foghlú, environmental scientist and woodland conservationist, calls for a more proactive approach in establishing upland forests, which he believes "are hostile" to woodland creation by the Forest Service's own rules:

> For upland farmers living outside protected areas, the Forest Service's own rules make afforestation equally difficult, Many of these rules are based on outdated views on a potential site's "production capacity" – this ignores the broad range on non-timber pub lic goods trees can provide.

> The official altitude limit for broadleaf trees in the west is just 120m. Sites with over 25% bare rock or sites proximate to the sea are also ruled out. Such conditions result in vast areas of land, where trees would happily grow, being disqualified from schemes. Recently introduced criteria relating to High Nature Value lands, breeding waders and shallow peats further narrow the scope of upland afforestation.

Ó Foghlú is in favour of protecting rare habitats and endangered species, but says:

> Viewed alone, each one can be resolutely defended but applied collectively, they function as an effective ban on upland woodland creation.

The virtual banning of afforestation on unenclosed upland sites in Ireland has destroyed any chance of recreating the primeval forests of Scots pine and oak on the uplands and lower slopes. This decision, which was carried out without consultation, has been typical of some of the blanket solutions taken in recent years that have alienated upland farmers and foresters. Projects such as Maam, Co. Galway and the Coronation Forest, Co. Wickow illustrate the huge wood and non-wood potential of creating – or recreating – upland forests, where they once flourished.

FUTURE SILVICULTURE SYSTEMS

Lesson one. The ways to propagate a tree are many.
Some take root on their own, with no one's help,
and put themselves about the place ...
Virgil, 29BC, from *Georgics,* translated by Peter Fallon, 2004

In *Georgics*, Virgil describes six ways to propagate trees. He begins with three types of natural regeneration or "Nature's way for each and every tree in woods and sacred groves to thrive and flourish". These are followed by three "other ways, found out by trial and error"; in other words where man intervenes to plant or coppice.

Clearcut rotation

Like most of Europe and North America, Irish foresters and forest owners rely on one method of establishing forests. All afforestation and most reforestation is carried out by manual or mechanical planting of trees, sourced from forest nurseries. Forest management, or silviculture, as practised in Ireland is straightforward: establish a crop; thin it every four or five years when it reaches "pole" stage and clearfell it at maturity.

This crop cycle resonates with farmers and foresters as it has a beginning, middle and rotation end. In reality it has no ending as no sooner is the crop harvested, the growing cycle begins again as reforestation is a mandatory requirement in Ireland. The difference between forest and agricultural management lies in time: agriculture is seasonal while forestry is generational or multi-generational, depending on tree species. Fast-growing conifers may be clearfelled by year 30; slower-growing conifers such as Scots pine may reach maturity at around 60 years, while broadleaves such as beech and oak are unlikely to reach final harvest until at least 100 years. The forest owners, with one eye on the short-term and one on the distant future, might sum up species selection as: spruce for us, pine for our children and oak for our children's children.

Clearcut rotation forestry has economic benefits for forest owners and the forestry and forest products sector. As it has clearly defined forest yield rotations, it allows accurate timber forecasting at local and national level, which is important in planning future timber processing, manufacturing, energy and related downstream industries. It provides opportunities for the forest owner to optimise earnings as clearfells can be delayed or brought forward, depending on market demand and log prices. Also, the grower can reduce rotations on exposed or unstable sites which may be subject to storm damage.

But it also has disadvantages, not least the ugly scar it leaves on the landscape after clearcutting and the risk of siltation of watercourses during large-scale felling. Although, the re-forested crop and associated vegetation quickly recolonise, the forest continuum is broken. It also results in an even-aged rather than a mixed-aged forest, although this is lessened as productive conifers can only be planted on 65% of forested areas, with the remainder carrying either native species or open biodiverse areas. This mix of productive conifers, broad-leaves and open spaces allows for a more organic expansion of forestry that is compatible with landscape and biodiversity values, although reducing the profitability of the forest.

Like most of Europe and North America, Irish foresters and forest owners rely on one method of establishing and managing forests. Forest management, or silviculture, as practised in Ireland, is straightforward: establish a crop; thin it every four or five years when it reaches "pole" stage and clearfell at maturity.

Clearcut is likely to remain as the preferred silvicultural system, especially for productive forests, located on wind vulnerable sites. In windfirm sites, there is a strong argument for continuous cover forestry (CCF), a silvicultural practice, sometimes referred to as close to nature forestry or irregular forest management. CCF maintains continuous woodland cover – "natures way" as Virgil said.

Continuous cover forestry (CCF)

I have visited CCF sites in Ireland and Europe, and a few common characteristics emerge that demonstrate its benefits as a worthwhile silvicultural system. Its main advantage lies in its ability to allow the emergence of a number of tree species and a rich ground flora. It works extremely well in pure broadleaf and coniferous stands, but its benefits are best demonstrated in Ireland in mixtures of conifers and broadleaves. What it provides is a forest continuum where the forest cycle ecosystem is kept intact, with harvesting and ground disturbance minimised. Based on forests I have seen in Ireland and overseas, especially in Belledonne Mountain near Grenoble, France and the primeval forest of Ravna vala southwest of Sarajevo, Bosnia and Herzegovina, the CCF argument is compelling, especially in maintaining and enhancing biodiversity. Amazingly, research carried out by Sarajevo University students in Ravna vala shows that the CCF-managed forest is richer in flora biodiversity than the untouched primeval forest it surrounds. There are biodiversity similarities between all CCF stands although they may differ in age and species composition. Cloragh near Ashford, Co. Wicklow is a predominantly coniferous forest but when I visited it in May 2022, I was struck by how greatly it differed from conventionally managed forests, which can be dark and have low ground vegetation cover.

The first impression you get when walking through Cloragh forest is the profusion of light. Even though it is mostly coniferous forest, shafts of sunlight permeate the canopy, providing a mixed mosaic of dappled colours on the verdant forest floor.

The second feeling is permanency. This multi-storeyed forest seems like it has existed for centuries. Majestic Douglas fir and Sitka spruce coexist with older Scots pine while below the canopy, diverse trees species of varying heights vie for room and light. Young western hemlock competes with larch, fir, spruce and pine while individual oak, birch and sweet chestnut bide their time in a nature-orchestrated forest. But then, Ballycurry is no ordinary forest. Liam Byrne, who manages the forest, acts as silvicultural conductor in directing the individual tree species' performances while ensuring continuity and permanence, which are the hallmarks of CCF. He allows the Scots pine and oak their own space as solo native performers while Japanese larch and rapidly regenerating western hemlock provide supporting roles but these are kept in check to allow the main species – Sitka spruce and Douglas fir – to play the lead roles. Like all good conductors, he is conscious of his

CONTINUOUS COVER

CCF is a silvicultural system that ensures permanent tree cover, thus avoiding clearfells. "It is a forest management system that integrates the environmental, social and economic benefits of forestry by maintaining the forest canopy and embracing natural regeneration to replace harvested trees," claims Pro Silva, the organisation that has been promoting CCF since it was founded in 2000. While the system is new to Ireland, it is practised throughout the world, especially in broadleaf forests, but it can be applied to pure broadleaves, pure conifers or mixtures. The modern European concept of CCF or "close to nature" silviculture was developed by foresters such as Karl Gayer, Adolph Gurnaud and Henry Biolley at the end of the nineteenth century, notably in Switzerland and Slovenia. Biolley perfected the jardinage single-tree selection system in Couvet in the Swiss Jura, where it is still practised. The jardinage or garden reference is a key to the system, which requires attention to detail whereby individual trees are selected almost with the same familiarity, as a gardener. "The single tree selection was required because plantation – clearcut – forests were vulnerable in certain locations to catastrophic landslides, avalanches, wind damage, insect damage and loss of soil fertility and structure," according to Paddy Purser, Pro-Silva Ireland.

Overleaf: Forest manager, Liam Byrne in a Sitka spruce dominated forest in Cloragh, Co. Wicklow where CCF has been practised since 2005. The once even-aged forest has transitioned to an uneven-aged forest. Naturally regenerated species also include Douglas fir, Norway spruce, Scots pine, oak and birch.

audience – and customers – so the lead parts are consigned to the tree species that generate a steady flow of logs to sawmills and a continuous income stream for the forest owner. "The forest that pays is the forest that stays," he says. "The main objective is to produce quality timber. The social and environmental benefits follow." While the forest seems ancient, it isn't that old at all. The dominant western North American natives – up to 40m tall – along with the middle and rapidly regenerating understorey bear no resemblance to the forest that existed here only a few decades ago.

CCF requires a different silvicultural approach to clearcut management. It needs continuous management to maximise its benefits. The costs are higher than conventional forestry but the economic dividends make it worthwhile. Tree removal requires extremely skilled harvester operators who understand the synergies involved in CCF. In this regard, Byrne relies on the expert harvesting skills of his brother Paul. While Douglas fir and Sitka are the major species, room is made for other species. such as Scots pine and oak. We walk past a well-formed oak which was allowed to emerge because a massive nearby 7m^3 Sitka was removed to make way for the oak, which will eventually form an upper-storey species. There are examples of other species, including birch and mountain ash being allowed the space to contribute to the forest ecosystem. He is conscious of the role these species play in improving species diversity and soil amelioration. "We are only learning now how these species interact with each other," he explains. "There are synergies happening here that we don't fully understand, especially in the root system." Because, so few trees are removed, there is no damage to the soil or other trees.

As a forester I would encourage CCF on all broadleaf forests and woodlands and coniferous and mixed-species plantations on windstable sites. I would also urge caution on a blanket drive towards CCF especially on wind-vulnerable sites. Pro Silva Ireland acknowledges the importance of soil and site factors when planning new CCF forests or transitioning from existing forests to continuous cover. Ideally CCF trees grow on to 40m top height, while trees that grow beyond 25m are vulnerable to windblow on many sites in Ireland, so clearcutting followed by replanting will still have a role to play.

Agroforestry

In deciding the forest type and management regime best suited to Irish climatic and site conditions, it is critical to keep an open mind, as no size fits all. There is room for both CCF and clearfell management, just as there is need for openness in other forest types such as agroforestry, which may be best suited for farmers who wish to farm and grow trees on the same plot of land (see page 27}. Ultimately, once environmental guidelines are observed, forest owners should not only have the autonomy on how they manage their forests but also where they wish to create the forests of the future.

Agroforestry is an alternative silvicultural system and habitat for livestock, while the wide spacing of trees also allows cropping for silage, hay and cereals.

A STRUCTURE TO DELIVER FUTURE FORESTS

*[To conserve the] forest, and to take from it what we need, we are
obliged to manage it.*
Colin Tudge, *The Secret Life of Trees: How They Live and Why
They Matter*, 2005

Since Ireland's Climate Action Plan 2019 identified forestry as
the most important land use to achieve carbon neutrality by
2050, there has been an expectation that afforestation would
need to dramatically increase to achieve 18% of the land area
within this timeframe. Unfortunately, this expectation has coincided with the lowest period of afforestation since the 1930s.
While the State has provided the basis for a major forestry programme for almost a century, it is now widely acknowledged
that a business-as-usual approach will not achieve net zero
which is central in national climate policy. COFORD – the government's forestry advisory body – estimates that annual afforestation needs to increase to 16,000ha while the University
of Galway study *The Economics of Afforestation and Management in Ireland: Future Prospects and Plans* maintains an 18,000-
ha programme is required.

The enormity of the challenge is apparent as Ireland's
afforestation programme has fallen to an average of 2,000ha
from 2020 to 2023. Stakeholders representing forestry companies, foresters, farmers and timber processors believe that
the current departmental structure which served forestry well
especially during the State's afforestation drive in the last century is no longer equipped to achieve the major challenges that
the sector requires. The failure of the current structure was
acknowledged in the 2014 Department of Agriculture, Food and
the Marine (DAFM) strategic plan *Forests Products and People –
Ireland's Forest Policy*, which recommended the setting up of a
task force:

> to consider the establishment of a stand-alone government body or
> agency which could have the responsibility of addressing the
> developmental and promotional issues of the forest sector.

The policy's authors were conscious that annual afforestation had fallen from 10,096ha to 6,100ha and that this trend
was likely to continue. However, the Department reneged on
its promise to examine the feasibility of an agency, while exas-
perated stakeholders watched annual afforestation in freefall
as planting dropped to levels even below those achieved in the
mid-1930s. Since 2007, stakeholders have called for the establishment of an independent State agency to oversee forestry
development in Ireland. Forestry is the only natural resource
in Ireland without such an agency to lead and promote it, unlike
ports and shipping (IMDO), food (Bord Bia), sea fisheries
(BIM), inland fisheries (IFI), marine research (Irish Marine Institute) and renewable energy (SEI). The importance of these
structures is acknowledged by Muiris MacCarthaigh:

State agencies have contributed much to Irish society and government and are an integral part of our system of public administration. Amongst other things, they enable the state to extend its capacity and respond in a flexible and responsive manner to citizens' needs.

In his study *The Economics of Afforestation and Management in Ireland*, Prof. Cathal O'Donoghue called for an ambitious annual afforestation programme of 18,000ha if Ireland is to halve its carbon emissions by 2030 and reach net zero emissions no later than 2050, but acknowledged that the current structure will not deliver a viable forestry programme:

> The present governance structure of the forest industry eco-system is itself fragmented with different state agencies having responsibility. There is also an overlap between policy and regulatory and development functions. Given the unique circumstances faced by the sector and the large societal benefits that the sector can deliver, there is a merit in exploring new governance structures such as establishing a Forestry Development Agency (FDA) to undertake a leadership role in developing the sector and to coordinate and deliver actions within the sector.

The establishment of an FDA would bring forestry in line with other natural resource sectors according to Michael Guilfoyle, who served as Assistant Secretary General in the then Department of Communications, Marine and Natural Resources (DCMN):

> Except for forestry, the promotion and development of these sectors, within overall Government policy, were statutorily delegated to proactive, highly professional bodies whose sole role was the furtherance of the contribution by these sectors, in consultation with a range of interest groups.

Guilfoyle's experience in DCMN when he was involved in setting up State agencies such as IMDO and the Marine Institute convinced him of their importance in developing their sectors and in their positive relationship with their parent government department. He rejects the idea that an FDA would be a threat to or in competition with the Forest Service – the forestry division in the Department of Agriculture, Food and the Marine (DAFM):

> Regarding competition, the exact opposite would be the case. FDA would be "under" and subject to the corporate governance of the Forest Service [in DAFM}. The FDA services and knowledge base would be available to the Forest Service and the Minister on an ongoing basis; its value would be as the professional body in constant touch with the sector, international developments, interest groups and with the policy challenges and threats to the sector.

The original call for an FDA by Guilfoyle, John McCarthy, director of None-so-Hardy Nurseries, the late Dr. Gerhardt Gallagher and this writer in in 2007, was based on the need for a cohesive approach to forest planning and development, as few sectors have a greater degree of interdependency than forestry. How the various links along the forest value chain perform and interact determine the viability of the sector as a whole. This includes the quality and productivity of our forests, the performance of downstream industries and how effectively the sector delivers a range of social and environmental benefits.

These are essential elements of the independent forestry agency's culture and mission that acknowledge multipurpose forestry as Ireland's key land use and wood resource in combating climate change. The challenge facing such an agency in creating a forest and wood culture is enormous. Acknowledging the environmental, economic and social interdependency of forestry is a prerequisite of the government when establishing this agency. Support from the government would also be an acknowledgement of the need for a partnership approach. which chimes with Edmund Burke's visionary social contract:

> Society is indeed a contract ... a partnership in all science; a partnership in all art; a partnership in every virtue, and in all perfection. As the ends of such a partnership cannot be obtained in many generations, it becomes a partnership not only between those who are living, but between those who are living, those who are dead and those who are to be born.

Burke's reflection on the social contract applies to forestry more than any other endeavour. A State-civil partnership is essential to sustain a forest continuum that has clear long-term objectives but flexibility to respond to changing environmental, social and economic circumstances. This requires an understanding of forests past and present in order to steer a pathway to creating forests of the future for "those who are to be born". A new approach, which an independent State forestry agency can provide, is vital in this new contract which caters for the needs of all its people.

REFERENCES

INTRODUCTION

Attenborough. 2019. "Our Planet: How to Save our Forests", *Our Planet* series, Netflix.

Boland. E. 2013. *Code*, Carcanet Press Ltd. Manchester.

Leskinen, P. 2018. "Recognising the role of forests in a sustainable European bioeconomy is fundamental", presentation at Society of Irish Foresters "Forestry as a Central Pillar in the Bioeconomy", conference, Society of Irish Foresters. Glenealy.

Sundström, K-H. 2019. "The Answer is Blowing in the Trees" *Evolution*, interview with Jan Lindroth, *Evolution* (Timber Technology Magazine, SKF). Gothenburg. https://evolution.skf.com/the-answer-is-blowing-in-the-trees/

Tudge, C. 2005. *The Secret Life of Trees: How They Live and Why They Matter*, Penguin Books Ltd., London.

1. FOREST, WOODLAND OR PLANTATION: DEFINING TREE COVERED LAND

Allaby, M. (ed.). 1992. *The Concise Oxford Dictionary of Botany, Oxford University Press.*

Dante Alighieri. 165-1321. *The Divine Comedy of Dante Alighieri*. Translated by Henry Wadsworth Longfellow, Project Gutenberg eBook. https://www.gutenberg.org/cache/epub/1004/pg1004.txt

Rackham, O. 2006. *Woodlands*, Harper Collins, London.

de Bhaldraithe, T. (ed.). 1959. *English-Irish Dictionary*. Oifig and tSolathair, Dublin.

Heaney, S. 1975. "Exposure", *North*, Faber & Faber Ltd. London.

Longfellow, H.W. 1988. Henry Wadsworth Longfellow: Selected Poems, Penguin, London. *(The Complete Poems of Henry Wadsworth Longfellow)*, originally published in 1859, Kindle edition, 2012.

Munro, A. 2005. "Chance", *Runaway*. Vintage, London.

Munro, A. 2010. "Wood", *Too Much Happiness*. Vintage, London.

O'Keeffe, J.G. (ed.). 1931. *Buile Suibhne*. Medieval and Modern Irish Series, Volume I, The Dublin Institute of Advanced Studies, Dublin.

Nelson, C. and W. Walsh. 1993. *Trees of Ireland – Native and Naturalised*. The Lilliput Press, Dublin.

Smyth, W.J. 2016. "The Greening of Ireland – Tenant Tree-Planting in the 18th and 19th Centuries", *Irish Forestry*, Dublin.

2. FORESTS PAST, FORESTS PRESENT: THE UNCERTAIN NARRATIVE OF IRISH FORESTRY

Carey, M. 2009. *If Trees Could Talk: Wicklow's Trees and Woodlands over Four Centuries*, COFORD, Dublin.

Carney, J. (ed.). 1967. "The Scribe in the Woods" *Medieval Irish Lyrics*, Berkeley: University of California Press, Oakland.

Carney, J. 1975. "The Invention of the Ogam Cipher", *Ériu*, Vol. 22, Royal Irish Academy, Dublin.

Cognisense. 2021. *Northern Ireland Forest Visitor Survey 2019*. Forest Service, Department of Agriculture, Environment and Rural Affairs, Belfast.

Crawford, H. 2023. *The Irish Forestry Society 1902-1923*, Society of Irish Foresters, Glenealy, Co. Wicklow.

Feehan, J. 2003. *Farming in Ireland: History, Heritage and Environment*, University College Dublin, Dublin.

Finneran, J. (ed.). 1983. *W.B. Yeats: The Poems*. Gill and Macmillan Ltd., Dublin.

Fitzpatrick. H.M. 1965. *The Forests of Ireland – An Account of the Forests of Ireland from Early Times until the Present Day*. Society of Irish Foresters, Dublin.

Hall, V. A. 1994. "Woodland Depletion in Ireland Over the Last Millennium." *Wood, Trees and Forests in Ireland, Proceedings of a Seminar Held on 22 and 23 February*. Royal Irish Academy, Dublin.

Hall, V. A. 1997. "The Development of the Landscape of Ireland Over the Last Two Thousand Years: Fresh Evidence from Historical and Pollen Analytical Studies." *Chronicon*, Vol I (Damian Bracken, ed.), University College Cork.

Heaney, S. 1972. *Wintering Out*, Faber & Faber Ltd. London.

Heaney, S. 1984. *Sweeney Astray*, Derry: Field Day Theatre Company, Derry and London: Faber and Faber.

Joyce, J. 1960. *Ulysses*, The Bodley Head Ltd., London. (First published in 1922 – extracts from Bodley Head edition.)

Joyce, P.W. 1995 (facsimile). *Irish Names of Places*. With a new introductory essay on the life of P.W. Joyce by Mainchín Seoighe. Three volumes. Éamonn de Búrca for Edmund Burke Publisher, Dublin.

Kearney, B. and R. O'Connor, R. 1993. *The Impact of Forestry on Rural Communities*. The Economic and Social Research Institute, Dublin.

Kelly, F. 2000. *Early Irish Farming*. Dublin: Dublin Institute for Advanced Studies. Reprinted by the School of Celtic Studies, Institute for Advanced Studies. Dublin.

Lucas, A.T. 2017. *The Sacred Trees of Ireland*. Society of Irish Foresters, Glenealy, Co. Wicklow. (Originally published by The Journal of the Cork Historical and Archaeological Society).

Mac an tSaoir, S. and J.R. Pilcher. 1995. *Wood, Trees and Forests in Ireland: Proceedings of a Seminar Held on 22 and 23 February*. Dublin: Royal Irish Academy, 1995.

Mac Coitir, N. 2003. *Irish Trees: Myths, Legends and Folklore*, The Collins Press, Cork.

MacNeill, M. (trans.). 1987. "The Scribe in the Woods", *An Irish Literature Reader: Poetry, Prose, Drama*, Syracuse University Press, New York.

Mitchell F. and F Ryan.1997. *Reading the Irish Landscape*, Town House and Country House, Dublin.

Neeson, E. 1991. *A History of Irish Forestry*, Lilliput Press Limited, Dublin.

O'Keeffe, J.G (translation). 1913. *Buile Suibhne*, Irish Text Society, Dublin.

Ó Rathaille, A. c1720. *O Rathaille,* translated by M. Hichael Hartnett, The Gallery Press, Oldcastle, Co. Meath.

Ó Túama, S. and T. Kinsella. 1981. *An Duanaire – Poems of the Dispossessed*, The Dolmen Press, Mountrath.

Pim. S. 1966. *The Wood and the Trees – A Biography of Augustine Henry.* Macdonald & Co. Ltd. London.

Rackham, O. 2006. *Woodlands*, Harper Collins: London.

Ryan M., V. Upton, C. O'Donoghue. 2015. Valuing the Ecosystem Services of Irish Forests, 2015. Teagasc.

Tree Register of Ireland. 2022. Tree Council of Ireland, Dublin.

Toner, G., M. Fomin, G. Bondarenko and T. Torma. 2007. *Electronic Dictionary of the Irish Language.* Web publication/site, Royal Irish Academy. http://www.dil.ie

3. FOREST PROTECTION

Barthlott, W., K. E. Linsenmair, K., S. Porembski. (eds.). 2009. *Biodiversity: Structure and Function* (Volume II, Encyclopedia of Life Support Systems). EOLSS Publishers Co. Ltd., Oxford.

Dearborn, K.D., C.A. Wallace, R. Patankar and J.L. Baltzer. 2021. "Permafrost thaw in boreal peatlands is rapidly altering forest community composition", *Journal of Ecology* (Vol. 109; Issue 3; 1452-1467), British Ecological Society, London. https://doi.org/10.1111/1365-2745.13569

Department of Agriculture, Food and the Marine. 2022. *Forest Statistics Ireland 2022*. Wexford.

European Commission. 2010. "Green Paper on Forest Protection and Information in the EU: Preparing forests for climate change" (COM/2010/0066 final). Brussels.

Food and Agriculture Organization of the United Nations. 2020. *Global Forest Resources Assessment. 2020 – Key Findings*. FAO, Rome. 2020, https://doi.org/10.4060/ca8753en

Gardiner, B., A. Schuck, M-J.O. Schelhaas, C. Orazio, K. Blennow and B. Nicoll. (eds.). 2013. *What Science Can Tell Us: Living with Storm Damage*, European Forest Institute, Joensuu.

Heaney, S. 2004. Foreword to *Devil's Glen; Sculpture in Woodland*. Magner, D. (ed). Sculpture in Woodland (Ireland) Ltd., Wicklow.

Ní Dhubháin, Á. and N. Farrelly. 2018. "Understanding and Managing Windthrow" (Silviculture/Management No. 23). *COFORD Connects*. COFORD, Department of Agriculture, Food and the Marine, Dublin. http://www.coford.ie/media/coford/content/publications/2018/SM23ManagingWindthrow160418.pdf

United Nations Environment Programme. 2020. *The State of the World's Forests 2020: Forests, Biodiversity and People*, Food and Agriculture Organization of the United Nations, Rome. https://doi.org/10.4060/ca8642en

United Nations Environment Programme. 2022. *Spreading like Wildfire – The Rising Threat of Extraordinary Landscape Fires*. A UNEP Rapid Response Assessment. Nairobi.

4. SUSTAINABLE FOREST MANAGEMENT

Attenborough, D. 2019. "Our Planet: How to Save our Forests", *Our Planet* series, Netflix.

Chamshama, S.A.O. and F.O.C. Nwonwu. 2004 "Forest plantations in Sub-Saharan Africa", A report prepared for the project "Lessons Learnt on Sustainable Forest Management in Africa". Joint publication by Royal Swedish Academy of Agriculture and Forestry (KSLA) and African Forest Research Network (AFORNET) at the African Academy of Sciences (AAS), and Food and Agriculture Organisation (FAO), UN, New York.

Department of Agriculture, Food and the Marine. 2022. *Forest Statistics Ireland 2022*. Johnstown Castle, Wexford.

Forest Service, Northern Ireland. 2019. Northern Ireland Forests Visitor Survey 2019, Department of Agriculture, Environment and Rural Affairs, Belfast. https://www.daerani.gov.uk/sites/ default/files/ publications/daera/181352_Northern%20Ireland%20Forest%20Survey%20Report%202019%20%20FINAL %20v9%2022.10.20.PDF

Forest Stewardship Council (FSC), 2023. The Future of Forests is in Our Hands, FSC, Bonn. https:// fsc.org/en

Merivale, W. 2023. "PEFC forest certification for woodland owners", *Forestry and Timber Yearbook* (D. Magner, ed.), Irish Timber Growers Association, Dalkey.

Ministerial Conference on the Protection of Forests in Europe - FOREST EUROPE. 2020. *State of Europe's Forest 2020*. (Coordinator Rastislav Raši), Bratislava. https://foresteurope.org/wp-content/uploads/2016/08/ SoEF_2020.pdf.

Programme for the Endorsement of Forest Certification (PEFC). 2023. "What is Sustainable Forest Management?", PEFC, Geneva. https://pefc.org

United Nations Forum on Forests. 2007. "Report of the seventh session (24 February 2006 and 16 to 27 April 2007)", Economic and Social Council, United Nations, New York. https://www.un.org/esa/ forests/wp-content /uploads/2013/09/E-2007-42-UNFF7Report.pdf

Yale Forest Forum. 2014. *The Global Forest Atlas*, Yale University, New Haven. https://globalforestatlas .yale.edu /conservation/forest-certification

5. WHY FORESTS?

Agricola, G. 1556. *De Re Metallica*. Translated by Herbert Clark Hoover and Lou Henry Hoover. New York: Dover Publications, 1986. Reprint of the 1950 reprint of the 1912 edition. https://www.gutenberg.org/files/38015/38015-h/38015-h.htm

Bengston, D.N.. 2018. "The Revolutionary Role of Wood in our Future", Forest Service, U.S. Department of Agriculture, arch and Development, USDA Forest Service., Washington DC. https://www.usda.gov/ media/blog/2018/01/05/revolutionary-role-wood-our-future

Botanical Society of Britain & Ireland. 2023. *Ireland's Changing Flora: A Summary of the Results of Plant Atlas 2020*. plantatlas2020.org

Carey, M., Hamilton, G., Poole, A. and Lawton, C. 2007. The Irish squirrel survey 2007. COFORD, Dublin.

COFORD. 2022. Forests and wood products, and their importance in climate change mitigation: A series of COFORD statements. COFORD, Dublin.

Cross, J.R. and K.D. Collins. 2017. *Management Guidelines for Ireland's Native Woodlands*, Jointly published by the National Parks and Wildlife Service and the Forest Service, Dublin.

Curtis, T.G.F. and H.N. McGough. 1988. *The Irish Red Data Book: No.1 Vascular Plants*, Wildlife Service Ireland, Dublin.

Daltun, E.. 2022. An Irish Atlantic Rainforest: A Personal Journey into the Magic of Rewilding, Hachette Books Ireland, Dublin.

D'Arcy. G. 1981. *The Guide to the Birds of Ireland*. Irish Wildlife Publications. Dublin.

D'Arcy. G. 2014. "I feel some clarification is required on my part" (letter). *Irish Farmers Journal*, Dublin.

Dowding, P. and L. Smith. *Forest Fungi in Ireland*. Dublin: COFORD – National Council for Forest Research and Development, 2008. http://www.coford.ie/media/coford/content/publications/ForestFungiinIreland 2008160919.pdf

Ellison, D., B. Muys and S. Wunder. 2020. "What role do forests play in the water cycle?" EFI, Joensuu.

European Commission. 2010. EU Green Paper "On Forest Protection and Information in the EU: Preparing forests for climate change." Brussels.

European Commission. 2021. *New EU Forest Strategy for 2030*, Communication from the Commission to the European Parliament, the Council, the European Economic and Social Committee and the Committee of the Regions, Brussels. https://ec.europa.eu/info/sites/default/files/communication-new-eu-forest-strategy-2030_with-annex_en.pdf

European Environment Agency. 2018. "Forests Utilisation", European Environment Agency, Copenhagen. https://www.eea.europa.eu/airs/2018/natural-capital/forest-utilisation#:~:text=The%20forest%20utilisation%20rate%20is%20the%20ratio%20between%20the%20annual,or%20timber%20rserve% 2C%20is%20stable.

Forest Service, Department of Agriculture, Food and the Marine, 2018. *Ireland's National Forest Inventory 2017: Results*. Wexford.

Forest Service. 2000-2002. Following suite of environment guidelines referenced, produced by the Forest Service, Department of Communications, Marine and Natural Resources, Dublin.
- "Forestry and Water Quality Guidelines", (Ryan. M.G. (ed.).
- "Forestry and Landscape Guidelines" (McCormack, A., and T. O'Leary (eds.).
- "Forest Harvesting and the Environment Guidelines" (Purser Tarleton Russell Ltd. (ed.).
- "Forest Biodiversity Guidelines" (Iremonger, S. (ed.).
- "Forest Protection Guidelines" (Anon.).

Friedlingstein, P., M.W. Jones, M. O'Sullivan, R. M. Andrew, D. C. E. Bakker, J. Hauck, C. Le Quéré, G. P. Peters, W. Peters, J. Pongratz, J, S. Sitch, JG. Canadell, P. Ciais, R.B. Jackson, S.R. Alin, P. Anthoni, N. Bates, M. Becker, N.Bellouin, L. Bopp, T.T.T. Chau, F. Chevallier, L.P Chini, M. Cronin, K.I. Currie, B. Decharme, L. Djeutchouang, X. Dou, W. Evans, R.A. Feely, L. Feng, T. Gasser, D. Gilfillan, T. Gkritzalis, G. Grassi, L. Gregor, N. Gruber, Ö. Gürses, I. Harris, R.A. Houghton, G.C. Hurtt, Y. Iida, T. Ilyina, I.T. Luijkx, A.K. Jain, S.D. Jones, E. Kato, D. Kennedy, K. Klein Goldewijk, J. Knauer, J.I. Korsbakken, J. I. Körtzinger, P. Landschützer, S.K. Lauvset, N. Lefèvre, S. Lienert, J. Liu, G. Marland, P.C. McGuire, J.R. Melton, D.R. Munro, D. R., J. E. M. S Nabel, S-I. Nakaoka, Y. Niwa, T. Ono, D. Pierrot, B. Poulter, G. Rehder, L. Resplandy, E. Robertson, C. Rödenbeck, T.M. Rosan, J. Schwinger, C. Schwingshackl, R. Séférian, A.J. Sutton, C. Sweeney, T. Tanhua, P. Tans, H. Tian, B. Tilbrook, F. Tubiello, G. van der Werf, N. Vuichard, C. Wada, R. Wanninkhof, A. Watson, D. Willis, A.J. Wiltshire, W. Yuan, C. Yue, X. Yue, S. Zaehle and J. Zeng, 2021. *Global Carbon Budget 2021*, Earth System Science Data, Göttingen. https://doi.org/10.5194/essd-2021-386, in review, 2021.

Farrelly, N. and Gallagher,G. 2015. "An analysis of the potential availability of land for afforestation in the Republic of Ireland", *Irish Forestry*, 120 – 138 (4).

Henry, A. 1966. Letter to Evelyn Gleeson (1899), from Pim. S. 1966. *The Wood and the Trees – A Biography of Augustine Henry.* Macdonald & Co. Ltd. London.

Irish Rare Birds Committee. 2014. The Irish List, BirdWatch Ireland, Greystones. https://www.npws.ie /sites/ default/files/files/2014IrishList(1).pdf

Irwin, S., D.L. Kelly, T.C. Kelly, F. J. G. Mitchell, .L. Coote, A. Oxbrough, M.W. Wilson, R.D. Martin, K. Moore, O. Sweeney, A. C. Dietzsch and J. O'Halloran. 2013. "Do Irish Forests Provide Habitat for Species of Conservation Concern?" *Biology and Environment: Proceedings of the Royal Irish Academy*, Royal Irish Academy, Dublin.

Irwin, S., S.M. Pedley, L. Coote, A.C. Dietzsch, M.W. Wilson, A. Oxbrough, O. Sweeney, K. M. Moore, R. Martin, D. L. Kelly, F.J.G. Mitchell, T.C. Kelly and J. O'Halloran. 2014. "The value of plantation forests for plant, invertebrate and bird diversity and the potential for cross-taxon surogacy", *Biodiversity and Conservation*. Springer, Berlin. https://doi.org/10.1007/s10531-014-0627-4

Iwata, Y., A. Ni Dhubhain, C. Burke, J. Brophy and D. Roddy. 2015.*Woodlands for Health*. (In association with Coillte, Mental Health Ireland, Mental Health Assoication, HSE, DAFM, UCD), Dublin.

Jackson, M.W., Ú. FitzPatrick, E. Cole, M. Jebb, D. McFerran, M. Sheehy Skeffington and M. Wright. 2016. *Ireland: Red List No. 10: Vascular Plants*, National Parks and Wildlife Service, Department of Arts, Heritage, Regional, Rural and Gaeltacht Affairs, Dublin.

Jeffree, M. (ed.). 2019. *Wood: Building the Bioeconomy*. The European Confederation of Woodworking Industries, CEI-Bois, Brussels.

Kelly, K. 2019. "Recommendations for Nature Friendly Forestry: Greening Irish Forestry", BirdWatch Ireland, Greystones.

Lawton, C., R. Hanniffy, V. Molloy, C. Guilfoyle, M. Stinson and E. Reilly. 2019. *All-Ireland Squirrel and Pine Marten Survey*, National Parks and Wildlife Service, Dublin.

Lowery, M. 1990. "Forestry – its Part in Rural Development", *The Right Trees in the Right Places*, (Charles Mollan and Michael Maloney, eds.), Royal Dublin Society, Dublin.

Lucey, J. and Y. Doris (eds) 2001. *Biodiversity In Ireland: A Review of Habitats and Species*, Environmental Protection Agency, Johnstown Castle Estate, Wexford.

Lusby, J., I. Corkery, S. McGuiness, D. Fernández-Bellon, L. Toal, D. Norriss, D. Breen, A. O'Donaill, D. Clarke, S. Irwin, J. L. Quinn and J. O'Halloran. 2017. "Breeding ecology and habitat selection of Merlin Falco *columbarius* in forested landscapes", British Trust for Ornithology, Norfolk. https:// www.ucc.ie/en/ media/research/planforbio/forestecology/Lusbyetal2017BS.pdf

Mark, J. J. 2015. "Newgrange", *World History Encyclopedia*. World History Publishing, Surrey. https:// www.worldhistory.org/Newgrange/ World Wildlife Fund. 217. "Forest Habitats – Overview". WWF, Washington. https://www.worldwildlife.org/habitats/forest-habitat.

Magner, D. *Stopping by Woods: A Guide to the Forests and Woodlands of Ireland*, The Lilliput Press, Dublin.

Marušáková and Sallmannshoferet M. 2019. *Human Health and Sustainable Forest Management*, Forest Europe (Liaison Unit), Bratislava.

McCarthy Keville O'Sullivan Ltd. 2018. "Draft Plan for Forests and Freshwater Pearl Mussel in Ireland" (SEA Environmental Report, prepared for the Forest Service), Dublin.

Mitchell-Jones, T. (adapted for Ireland by Fahy, O. and Marnell, F.). 2009. "Bats and Forestry" (based on EUROBATS publication *Bats and Forestry*), National Parks & Wildlife Service, Dublin. https://www .npws.ie/sites/ default/files/publications/pdf/Forestry_leaflet%5B1%5D.pdf.

Murphy, G. 2024. "Reintroductions and recent bird arrivals", personal correspondence, Limerick.

National Biodiversity Data Centre. 2020. "Biodiversity Inventory by Taxonomic Groups". Carriganore, Co. Waterford. https://biodiversityireland.ie/taxonomic-groups/#archive.

National Parks & Wildlife Service. 2012. "Birds: Bird Conservation and Monitoring", National Parks & Wildlife Service, Dublin. https://www.npws.ie/research-projects/animal-species/birds

National Biodiversity Data Centre:A Heritage Council Project. 2021 (ongoing). "Vascular Plants" A Heritage Council Project. https://www.biodiversityireland.ie/projects/biodiversity-inventory/taxonomic-groups/ vascular-plants.

Neff, J. 1996." Biodiversity in Ireland: a review of species diversity in the Irish flora", Heritage Policy Unit, Department of Arts, Culture and the Gaeltacht, Dublin.

Nisbet, T. 2014. "The Role of Productive Woodlands in Water Management". Confor and Forest Research, Edinburgh.

O'Callaghan, C.J., S.Irwin, K. A. Byrne, J. O'Halloran. 2016. "The role of planted forests in the provision of habitat: an Irish perspective". *Biodiverity and Conservation*. Springer Science+Business Media Dordrech. https://www.ucc.ie/en/media/research/planforbio/pdfs/publications/OCallaghanetal 2016.pdf

O'Halloran, J., P.M. Walsh, P.S. Giller, T.C. Kelly. 2002. "Forestry and Bird Diversity in Ireland: A Management and Planning Guide", COFORD, Dublin.

O'Kelly, M. J. 1982. *Newgrange: Archaeology, Art and Legend*. Thames & Hudson, London.

Parnell, J. and T. Curtis. 2012. *Webb's An Irish Flora*, Cork University Press, Cork.

Purcell, P. 1996. "*Biodiversity in Ireland: an inventory of biological diversity on a taxonomic basis*". Report submitted to the Heritage Policy Unit, Department of the Arts, Culture and the Gaeltacht, Dublin.

Rappold, G. (co-ordinator). 2007. T*he Austrian Forest Programme*, Republic of Austria, Federal Ministry of Agriculture, Forestry, Environment and Water Management, Vienna.

Rivers, M., E. Beech, I. Bazos, F. Boguni , A. Buira, D. Cakovi , A. Carapeto, A. Carta, B. Cornier, G. Fenu, F. Fernandes, P. Fraga i Arguimbau, P. Garcia-Murillo, M. Lepší, V. Matevski, F. Medina, M. Menezes de Sequeira, N. Meyer, V. Mikoláš, C. Montagnani, T. Monteiro-Henriques, J. Naranjo-Suárez, S. Orsenigo, A. Petrova, A. Reyes-Betancort, T. Rich, P. Harald Salvesen, I. Santana-López, S. Scholz, A. Sennikov, L. Shuka, L. F. Silva, P. Thomas, A. Troia, J. .L.Villar, and D. Allen. 2019. *European Red List of Trees*, IUCN (International Union for Conservation of Nature), Cambridge and Brussels.

Scannell, M.J.P. and D.M. Synnott. 1987. *Census Catalogue of the Flora of Ireland*. Clar de Phlandaí na hÉireann. 2nd. edition. Stationery Office, Dublin.

Scannell, M.J.P. and D.M. Synnott. 1990. "Records for the Census Catalogue of the Flora of Ireland in the Herbarium, National Botanic Gardens, Glasnevin", National Botanic Gardens. Dublin.

Smith, G., T. Gittings, M. Wilson, L. French, A. Oxbrough, S. O'Donoghue, J. Pithon, V. O'Donnell, A-M. McKee, S. Iremonger, J. O'Halloran, D. Kelly, F. Mitchell, P. Giller and Tom Kelly. 2005. "Assessment of Biodiversity at Different Stages of the Forest Cycle", BIOFOREST Project, COFORD, Dublin.

Stroh, P.A., K.J.Walker, T.A. Humphrey, O.L. Pescott and R.J. Burkmar. 2023. Plant Atlas 2020. *Mapping Changes in the Distribution of the British and Irish Flora*. 2 Volumes. Princeton University Press, Princeton.

van Romunde, R. 2020. *Global Timber Outlook 2020: An In-Depth Report on the Global Forestry Industry*, Gresham House, London. https://greshamhouse.com/global-timber-outlook/

Verstraeten, W.W., B. Muys, J. Feyen, F. Veroustraete, M. Minnaert, L. Meiresonne, A. De Schrijver. 2005. "Comparative analysis of the actual evapotranspiration of Flemish forest and cropland, using the soil water balance model", *WAVE. Hydrology and Earth System Sciences*, Munich.

Walker, K.J., Stroh, P.A., Pescott, O.L., Humphrey, T.A. & Roy, D.B. 2023. The changing flora of Britain. Durham: Botanical Society of Britain and Ireland.

Webb, D. A. 1983. "The Flora of Ireland in its European Context", *Journal of Life Science, Life Science* (vol 1983), The Boyle Medal Discourse 1982), Royal Dublin Society, Dublin. https://www.botanic-gardens.ie/wp-content/uploads/2018/06/David-A.-Webb.-1984-The-flora-of-Ireland-in-its-European-context.-J.-Life-Sci.-Roy.-Dublin-Soc.-143-160.pdf.

White, J. (John), J. (Jill) White and S. Max Walters. 2005. *Trees: A Field Guide to the Trees of Britain and Northern Europe*, Oxford University Press, Oxford.

Wilson, M.W., T. Gittings, J. Pithon, T.C. Kelly, S. Irwin, S. and J. O'Halloran. 2012. "Bird diversity of afforestation habitats in Ireland: current trends and likely impacts, *Biology and Environment: Proceedings of the Royal Irish Academy*, Dublin. http://www.coford.ie/media/coford/content/researchprogramme/landavailabiity/presentations/Wilsonetal2012BiolEnv280116.pdf

Whelan, D. (ed.), Á. Ní Dhubháin, R.Moloney, K. Black, E. Hendrick, T. Kent, E. O'Driscoll, S. Irwin and M. Cregan. 2014. "Irish Forests and Biodiversity", *Forestry 2030*, COFORD, Dublin.

6. WHY WOOD?

Architecture 2030. 2021. "Why the built environment" (Material sourced from Global ABC, Global Status Reports 2018 and 2021, EIA), Santa Fe. https://architecture2030.org/why-the-building-sector/

Boland, D. and A. O'Sullivan. 1997 'An early medieval wooden bridge at Clonmacnoise', In F.J.G. Mitchell and C. Delaney (eds.), *The Quaternary of the Irish Midlands*. Field Guide 21, Irish Association for Quaternary Studies, Dublin, 14-21.

BM Trada. 2022. "The Advantages of Timber Frame Construction". https://www.bmtrada.com/resources/the-advantages-of-building-in-timber-frame", BM Trada, High Wycombe. https://www. bmtrada .com /resources/the-advantages-of-building-in-timber-frame

Carey, J. *A Little History of Poetry*. London: Yale University Press, 2020.

Chang, P-C. and A. Swenson. 2020. "Construction", *Encyclopedia Britannica*, Chicago. https://www.britannica.com/technology/construction. Accessed 14 April 2022.

Clark, K.1969. *Civilisation*, BBC – John Murray, London.

Coillte. 2022. *A Greener Future for All: Strategic Vision for our Forests.* Coillte, Newtownmountkennedy, Co. Wicklow.

Coles, J.M., S.V.E. Heal, B.J. Orme. 2014. "Use and Character of Wood in Prehistoric Britain and Ireland", Proceedings of the Prehistoric Society, London.

Denmark.dk.2022. "Green thinking: Pioneers in clean energy", Ministry of Foreign Affairs of Denmark, https://denmark.dk/innovation-and-design/clean-energy

Dennehy, K. 2014. "Using more wood for construction can slash global reliance on fossil fuels", (Inter view with Chadwick Dearing Oliver, co author of "Carbon, Fossil Fuel, and Biodiversity Mitigation With Wood and Forests"), *Yale News*, Yale University. https://news.yale.edu/2014/03/31/using-more-wood-construction-can-slash-global-reliance-fossil-fuels

Ennos, R. 2021. *The Wood Age: How One Material Shaped the Whole of Human History*, William Collins, London.

Forest Research, Forestry Commission. 2022. "Woodland Statistics [UK], *Forestry Statistics 2021*, Edinburgh. www.forestresearch.gov.uk/statistics/

Government of Ireland. 2021. *Climate Action Plan: Securing Our Future*, Dublin.

Hanley, K. 2009. "Reconstructing prehistoric and historic settlement in County Cork". Dining and Dwelling, National Roads Authority, Dublin.

Harte, A.M. 2017. "Commercialisation of Irish Cross-Laminated Timber" Final Report to COFORD, Department of Agriculture, Food and the Marine (Ireland), NUI Galway.

Harte, A.M., B. Robinson, A. Macilwraith, M. Jacob,2017. *Timber in Multi-storey Construction*. COFORD, Department of Agriculture, Food and the Marine, Dublin. http://www.coford.ie/media/coford/content/publications/projectreports/TimberMultiStoreyConstruction310717.pdf

Hughes, R. 1997. *American Visions: The Epic History of Art in America*. The Harvill Press, London.

Jacob, M., J. Harrington, B. Robinson. 2018. *The Structural Use of Timber - Handbook for Eurocode 5: Part 1-1*. COFORD, Department of Agriculture, Food and the Marine, Dublin.

Kofman, P.D. and E. Hendrick. 2021. *Wood as a Fuel*, Volume 1 Fundamentals and Standards, Wood Fuel Book Partnership, Dublin.

Kordik, H. 2012. "Austria's Successful Rise in Woody Biomass", *New Austrian*, Washington. https://www.austrianinformation.org/summer-2012/2012/8/21/austrias-successful-rise-in-woody-biomass.html

Levey, M. 1996. *Florence: A Portrait*, Harvard University Press, Cambridge, Mass.

Lindroth, J. 2019. "The Answer is Blowing in the Trees" (Karl-Henrik Sundström interview), *Evolution*, https://evolution.skf.com/the-answer-is-blowing-in-the-trees/

Moffett, M., M. W. Fazio, L. Wodehouse. 2008. *A World History of Architecture*. Laurence King Publishing, London.

Natural Resources Canada. 2016. "Wood products: Everywhere for everyone." Government of Canada, Vancouver, British Columbia. https://www.nrcan.gc.ca/our-natural-resources/forests/industry-and-trade/forest-products-applications/non-timber-forest-products/wood-products-everywhere-for-everyone/

Natural Resources Canada. 2020. "Canada Investing in the Use of Wood Products for Automotive Applications", Government of Canada, Vancouver, British Columbia. https://www.canada.ca/en/natural-resources-canada/news/2020/10/canada-investing-in-the-use-of-wood-products-for-automotive-applications.html

O'Connor, C. (foreword). 2020. *Wood Awards Ireland* (ed. Magner, D) Wood Marketing Federation-Forest Industries Ireland, Dublin.

Oliver, C.D., N.T. Nassar, B. R. Lippke and J.B. McCarter. 2014 "Carbon, Fossil Fuel, and Biodiversity Mitigation With Wood and Forests", *Journal of Sustainable Forestry*, Taylor & Francis, London. https://doi.org/10.1080/10549811.2013.839386

O'Toole, D. 2021. " Impact of Increased Use of Timber in Construction". Forest Industries Ireland, Dublin.

Philip, J. 2017. *100 Contemporary Wood Buildings*, Taschen, Cologne

Project Ireland 2040: National Planning Framework. 2018. Department of Housing, Planning and Local Government, Dublin.

Ritchie, H. and M. Roser. 2018. "Plastic Pollution", *Our World in Data* (on line). https://ourworldindata.org/plastic-pollution

Sathre, R. and J. O'Connor. 2010. *A Synthesis of Research on Wood Products and Greenhouse Gas Impacts*, Tehnical Report No. TR-19R, 2nd Edition, Vamcouver B.C. https://www.canfor.com/docs/why-wood/tr19-complete-pub-web.pdf

Sathre, R. and J. O'Connor. 2010. "Meta-analysis of greenhouse gas displacement factors of wood product substitution", *Environmenal Science and Policy*, Science Direct, https://rogersathre.com/Sathre&OConnor_2010_wood_substitution_meta-analysis.pdf

Scarre, C. (ed.). 2018. *The Human Past: World Prehistory and the Development of Human Societies*. Thames and Hudson, London.

Suhonen, T. and T. Amberla (eds.). 2015. "Paper and paperboard market: Demand is forecast to grow by nearly a fifth by 2030", Pöyry Management Consulting, Vantaa.

Timpany Archaeology Institute—University of the Highlands and Islands, Orkney College, Kirkwall, Scotland, UKCorrespondencescott.timpany@uhi.ac.uk, S., A. Crone AOC Archaeology Group, Loanhead, Scotland, UK, D. Hamilton Scottish Universities Environmental Research Centre (SUERC), Radiocarbon Laboratory, University of Glasgow, East Kilbride, Scotland, UKhttp://orcid.org/0000-0003-4019-3823 and M. Sharpe. 2017. "Revealed by Waves: A Stratigraphic, Palaeoecological, and Dendrochronological Investigation of a Prehistoric Oak Timber and Intertidal Peats, Bay of Ireland, West Mainland, Orkney", *The Journal of Island and Coastal Archaeology*, Taylor & Francis, Inc., Philadelphia. https://www.tandfonline.com/doi/full/10.1080/15564894.2017.1284960

Tiseo, I. 2022. "Plastic waste worldwide - statistics and facts", Statista, New York. https://www.statista.com/topics/5401/global-plastic-waste/#dossierKeyfigures

Tudge, C. 2005. *The Secret Life of Trees: How They Live and Why They Matter*, Penguin Books Ltd., London.

Van der Lugt, P. and A. Harsta. 2020. *Tomorrow's Timber: Towards the Next Timber Revolution*, Material District, Naarden.

Verkerk, P.J., M. Hassegawa, J. Van Brusselen, M. Cramm, X. Chen, X., Y. I. Maximo, M. Koç, M. Lovri , and Y.T. Tegegne, 2021. "The role of forest products in the global bioeconomy – Enabling substitution by wood-based products and contributing to the Sustainable Development Goals." Rome. (FAO on behalf of the Advisory Committee on Sustainable Forest-based Industries (ACSFI))

Wormslev, E.C., J.L. Pedersen, C. Eriksen, R.Bugge, N. Skou, C. Tang, T. Liengaard, R.S. Hansen, J.M. Eberhardt, M.K. Rasch, J. Höglund, R. B. Englund, J. Sandquist, B.M. Güell, J.J.K. Haug, P. Luoma, T. Pursula and M. Bröckl. 2016. "Sustainable jet fuel for aviation: Nordic perpectives on the use of advanced sustainable jet fuel for aviation", Nordic Council of Ministers, TemaNord. Copenhagen.

7. THE RIGHT TREES IN THE RIGHT PLACES FOR THE RIGHT REASONS

Fitzpatrick, H.M. (ed). *The Forests of Ireland; An Account of the Forests of Ireland from Early Times to the Present Day*. Wicklow: The Society of Irish Foresters, 1965.

Forestry Act 2014. (enacted 2017) The Forestry Act 1988, the Forestry (Amendment) Act 2009 and Forestry Act 2014 cited as the Forestry Acts 1988 to 2014 and construed as one Act, Government Publications (electronic Irish Statute Book (eISB)).

Kavanagh, P. 2003. *Patrick Kavanagh: Collected Poems* (A. Quinn, ed.) Penguin Modern Classics, London.

Neeson, Eoin.1991. *A History of Irish Forestry*, The Lilliput Press, Dublin

OCarroll, N. 2004. *Forestry in Ireland: A Concise History*, COFORD, Dublin.

Pim. S. 1966. *The Wood and the Trees – A Biography of Augustine Henry*. Macdonald & Co. Ltd. London.

Virgil. 29BC. *The Georgics*, translated by J. Rhoades (1881). *Project Gutenberg eBook (updated 2021)*. https://www.gutenberg.org/files/232/232-h/232-h.htm

8. WHAT TREES? SPECIES THAT ADAPT TO IRISH SOIL, SITE AND CLIMATIC CONDITIONS

Anderson, L. 2005. "The ecology and biodiversity value of sycamore (*acer pseudoplatanus L*) with particular reference to Great Britain", *Scottish Forestry*, (59,3), Royal Scottish Forestry Society, Edinburgh.

Boland. E. j1822005. *New Collected Poems*. Carcanet Press Ltd. Manchester.

Botanical Society of Britain and Ireland (BSBI). 2020. Ireland's Changing Flora: *A Summary of the Results of Plant Atlas 2020*. https://plantatlas2020.org/sites/default/files/bsbi_data/home/BSBI%20Plant%20Atlas%2020 20%20summary%20report%20Ireland%20WEB.pdf. From P. A. Stroh, K. J. Walker, T. A. Humphrey, O. L. Pescott, and R. J. Burkmar. 2020. Plant Atlas 2020: Mapping Changes in the Distribution of the British and Irish Flora, Princeton University Press, Oxford.

Buckley, Y. 2023. "Managing aliens will improve our quality of life", *Irish Times*, Dublin.

Cross, J.R. and K.D. Collins. 2017. *Management Guidelines for Ireland's Native Woodlands*, Jointly published by the National Parks and Wildlife Service and the Forest Service, Dublin.

Elwes, Henry John and Henry, Augustine. 1906-1913. *The Trees of Great Britain and Ireland*, self published in seven volumes and index by Elwes, Edinburgh. Facsimile edition (2012) published by The Society of Irish Foresters, Wicklow.

Fitzpatrick, H.M. (ed). *The Forests of Ireland; An Account of the Forests of Ireland from Early Times to the Present Day*. Wicklow: The Society of Irish Foresters, 1965.

Forest Research. 2022. "Hybrid larch". Note in Forest Research, Forestry Commission (FC) Great Britain. https://www.forestresearch.gov.uk/tools-and-resources/tree-species-database/hybrid

Horgan, T., M. Keane, R. McCarthy, M. Lally and D.Thompson. 2003. *A Guide to Forest Tree Species Selection and Silviculture in Ireland* (ed, Joe O'Carroll), COFORD, Dublin.

Huss, J., P.M. Joyce, R. MacCarthy, J. Fennessy. 2016. *Broadleaf Forestry in Ireland*, COFORD, Department of Agriculture, Food and the Marine, Dublin.

Irish Green Building Council (IGBC). 2022. *Building a Zero Carbon Ireland, Dublin – A Roadmap to Decarbonise Ireland's Built Environment Across its Whole Life Cycle*. Dublin.

Johnson, H. 1973. *The International Book of Trees: A Guide and Tribute to the Trees of our Forests and Gardens*, Mitchell Beazley Limited, London.

Joyce, J. 1960. *Ulysses*, The Bodley Head Ltd., London. (First published in 1922 – extracts from Bodley Head edition.)

Joyce, P. and N. OCarroll. 2002. *Sitka Spruce in Ireland*, COFORD, Department of Agriculture, Food and the Marine, Dublin.

Joyce, P.M., J. Huss, R. McCarthy, A. Pfeifer and E. Hendrick. 1998. *Growing Broadleaves: Silvicultural Guidelines for Ash, Sycamore, Wild Cherry, Beech and Oak in Ireland*, COFORD, Dublin.

Magner, D. (ed.). 2020. Wood Awards Ireland. Wood Marketing Federation, Wicklow.

O'Sullivan Beare, D.P. 2009. The Natural History of Ireland : included in Book One of the Zoilomastix of Don Philip O'Sullivan Beare (c1625), translated and edited by O'Sullivan, D.C.), Cork University Press, Cork.

Roche, J.R. 2019. "Recent findings on the native status and vegetation ecology of Scots pine in Ireland and their implications for forestry policy and management." *Irish Forestry*, Vol. 76; Nos. 1&2. Glenealy, Co. Wicklow.

Mc Loughlin, J. 2016. "Trees and woodland names in Irish placenames", *Irish Forestry*, Vol 73 (239 – 257), Dublin.

Sundström. Erik. 1996. "Root architecture – a look into the underworld". *Making Headway with Forest Research*, Coillte Research & Development, Newtownmountkennedy, Co. Wicklow.

Tittensor. R. 2016. *Shades of Green: An Environmental and Cultural History of Sitka Spruce*, Oxbow Books, Oxford.

Tree Council of Ireland. 2005. *Champion Trees: A Selection of Ireland's Great Trees*, Tree Register of Ireland, Tree Council of Ireland, Dublin.

White, J. (John), J. (Jill) White and S. Max Walters. 2005. *Trees: A Field Guide to the Trees of Britain and Northern Europe*, Oxford University Press, Oxford.

9 OUR FUTURE FORESTS

Bail, M. 1998. *Eucalyptus*, Harvill Panther, London.

Burke, E. 1790. *Reflections on the Revolution in France*. Many editions including Penguin, 1982 and Oxford University Press, 1999.

COFORD. 2022. *Forests and wood products, and their importance in climate change mitigation: A series of COFORD statements*. COFORD, Dublin.

Coillte. 2022. *Strategic Vision for Our Future Forest Estate*, Coillte, Newtownmountkennedy, Co. Wicklow. Department of Agriculture, Food and the Marine. 2022. Public Attitudes Survey on Forestry (November / December 2021), Behaviour & Attitudes (B&A), Dublin.

European Forest Institute. 2018. "Expanding knowledge of community-based forestry in Europe". Joensuu. https://medforest.net/2018/11/15/expanding-knowledge-on-community-based-forestry-in-europe/

Farrelly, N. and Gallagher,G. 2015. "An analysis of the potential availability of land for afforestation in the Republic of Ireland, *Irish Forestry*, 120 – 138 (4).

Fitzgerald, John. 2023. "The Importance of Irish Woodland", *Forestry & Timber Yearbook 2024* (ed. Donal Magner), Irish Timber Growers Association, Dalkey.

Fleming, Michael. 2013. "Premiums and planting marginal land - key issues", *Irish Farmers Journal*, Dublin.

Gallagher, G., D. Magner , P. Phillips, C. O Carroll, P. O'Sullivan, P. Seppänen, D. Styles. 2022. *Development of a Carbon Farming Framework for Ireland*, Society of Irish Foresters, Glenealy, Co. Wicklow.

Irwin, Rachel, Ní Dhubháin, Áine and Short, Ian. 2022. "Irish dairy and drystock farmers' attitudes and perceptions to planting trees and adopting agroforestry practices on their land" (joint UCD-Teagasc study), *Environmental Challenges*, 2022, 9, 100636. doi: https://doi.org/10.1016/j.envc.2022.100636.

MacCarthaigh, Muiris. 2010. "National non-commercial State Agencies in Ireland", State of the Public Service Series - Research Paper No. 1, Institute of Public Administration, Dublin.

Magner, D. 2015. Evaluation of the RDS-Forest Service Irish Forestry Awards (submision to RDS, unpublished Dublin.

Magner, D. 2018. "Proposal for a Forestry Development Agency" (interview with M. Guilfoyle), 2018. *Irish Farmers Journal*, Dublin.

O'Donoghue, Cathal. 2022. "The Economics of Afforestation and Management in Ireland: Future Prospects and Plans", University of Galway, Biorbic Sfi Research Centre, Auxilia Group, Naas.

Virgil. 29BC. *Georgics*. Translated by Peter Fallon (2004), Oxford University Press, Oxford.

GLOSSARY

A

abiotic Non-living or composed of non-living elements (minerals, topography, geology, etc.). See biotic.
abiotic damage to trees Refers to damage caused by the following agent types: climate (climate variables such as exposure, frost and drought); nutrient deficiency; and anthropogenic factors (human causes).
acidic (soil) See pH.
afforestation Tree planting on bare land or land with previous crop other than trees or "artificial establishment by planting or seeding of forest on a non-forest area (e.g. agricultural or other land)," according to the European Forest Institute definition.
alkaline (soil) See pH.
angiosperms Flowering plants with seeds that are fully enclosed by fruits – broadleaf trees are classed as angiosperms. Angiosperms and gymnosperms are the two major groups of vascular seed plants.
apical dominance Growth concentrated on the leader – or leading shoot – which tends to produce a straight stem and conical crown, mainly in conifers.
aquatic zone A seasonal river or lake shore.
Area of Scientific Interest (ASI) Protected area that makes a major contribution to the conservation of a valuable natural site. See Natural Heritage Area (NHA).
auxins Hormones positioned at the growing points of stems and roots which regulate growth.

B

basal area Cross-sectional area of a tree measured at 1.3m diameter breast height (DBH) – usually measured in area of trees per hectare or in a specific area, expressed in square metres.
biodiversity (biological diversity) Term used to describe all aspects of biological diversity within an ecosystem such as genetic diversity, species diversity – including diversity within and between species – and landscape diversity.
biofuels Fuels derived from organic materials by biological processes, including wood biomass.
biotic Living components of an ecosystem as opposed to non-living abiotic.
biotic damage to trees Refers to the following living agents that cause damage to trees: disease; insect pests; animal; competing vegetation; and harvesting.
broadleaf trees or broadleaves Diffusely branched trees with broad, flat leaves, usually deciduous – such as chestnut, oak, ash, beech, birch and alder – but not always – such as holly, arbutus and eucalyptus.
bryophyte A division of the plant kingdom containing small, rootless non-vascular plants such as mosses.

C

calcareous soil Soil containing sufficient calcium carbonate to effervesce visibly when treated with 10% hydrochloric acid.
canker Disease caused by fungal or bacterial infection, usually resulting in bleeding from the tree bark.
canopy The mass of foliage and branches formed collectively by the crowns of trees.
carbon sequestration The absorption of carbon by plants during the process of photosynthesis and its storage, in the case of trees, until the timber decays or is destroyed.
certification A system that verifies that forests and woodlands are managed according to principles of sustainable forest management (SFM). It proves that these woods have been independently inspected and evaluated according to strict environmental, social and economic principles and criteria as agreed by FSC, PEFC or other recognised accredited body.
certification scheme A market oriented, voluntary, environmental scheme, designed to certify that forests are managed on a sustainable basis.
chain of custody certification Awarded to timber processors and manufacturers or others in the wood chain whose products can be traced to sustainably managed forests. Businesses with chain of custody certification

can stamp their products with the FSC, PEFC or logo of other recognised accredited certifying body.

clearcutting systems Silviculture management system applied in countries such as Ireland, UK, New Zealand, western North America, Sweden, and Finland. The crop is established by artificial generation. They are usually even aged crops that are thinned and eventually clearfelled. They are regenerated artificially although second and subsequent crops can be naturally regenerated.

clearfell coupe Clearfell area

clearfell Total tree felling in an area where trees have reached rotation end.

climax species The species in the final stage of plant succession – after pioneer species – which reaches a state of equilibrium with the environment. See also pioneer species.

close to nature forestry (or silviculture) See continuous cover forestry.

conifers Trees and shrubs that have needle-like leaves and bear cones. They are usually, but not always, evergreen. See also gymnosperms.

coppice Trees felled close to the ground so as to produce shoots or fresh growth from the resulting stools, giving rise to successive crops of trees.

cross-laminated timber (CLT) Engineered wood, manufactured by gluing together several layers of timber boards, with successive layers glued at right angles.

current annual increment (CAI) The volume increment of a crop of trees during the current year.

D

deciduous trees Temperate and boreal trees that shed all their leaves or needles in the autumn. Most broadleaves are deciduous – exceptions include holly, eucalyptus and arbutus – whereas only some conifers are deciduous – including larch and dawn redwood.

dominant trees The tallest and most vigorous trees in a forest block or stand.

dormancy The period in a plant's life when growth is suspended. Described by Michael Allaby as a resting period with reduced metabolic rate. This is found in non-germinating seeds and non-growing buds.

E

earlywood Thin-walled cells with large spaces (lumina) within the cell walls, formed in the annual ring at the commencement of growth in spring.

economic rotation The age at which the financial return is at its maximum.

ecosystem A community of plants and animals (including humans) interacting with each other and the forces of nature. Balanced ecosystems are stable when considered over the long term (hundreds of years, in the case of woodland).

enclosed land Land which is surrounded by man-made boundaries – hedgerows and walls in Ireland – and which was improved for agricultural use by cultivation and/or fertilisation. See also unenclosed land.

environmental guidelines Water quality, archaeology, landscape, biodiversity and harvesting guidelines of the Forest Service.

Environmental Impact Statement (EIS) A legally structured document describing the impacts of a development project on all aspects of the environment.

epicormic branches Branchlets originating from adventitious buds on the stem. They are found on most broadleaved species especially oak, poplars and Spanish chestnut. Often encouraged by an over heavy thinning.

eutrophication The process of nutrient enrichment in water ecosystems.

exotic species Also referred to as exotics, these are species that have been established by humans in places outside their native habitats.

exotic tree and plant species Generally refers to tree and plant species that have been deliberately or accidentally introduced since the nineteenth century as opposed to naturalised but introduced species.

F

final crop The trees which remain after successive thinnings and are finally felled at maturity.

flushed site A site with considerable enrichment with nutrients from flush water, as indicated by the presence and vigour of tufted hair grass, purple moor grass, soft and bog-rush species.

flushing The commencement of growth of a plant above ground, characterised by the swelling and bursting of buds.

forest biomass Total forest organic material including stems, leaves and roots.

forest reproductive material Seeds, plants or clones used for forestry purposes.

Forest Stewardship Council (FSC) An international, non-profit making association and a recognised accredited body which provides certification for forests and woodlands.

forest vitality The ability of the forest to endure and perform its functions.

forestry The management of predominantly tree covered land (woodland), whether in large tracts (generally called forests) or smaller units (known by a variety of terms such as woods, copses and shelterbelts).

forking Usually occurs when leader of terminal shoot is damaged. Double or multiple leaders usually replace the leader. This can be corrected by shaping or removing one or more of the forked branches to allow a leader to form. Forking of broadleaves should be corrected between years two and four.

frost damage Damage to the soft tissues of trees by cold temperatures, which can occur in the nursery and in young plantations. Trees are most vulnerable when freshly flushed in late spring, or early summer, and again in early autumn, prior to hardening off.

frost hollows Low-lying concave areas where cold air collects, causing damage to new shoots of trees.

fungi A group of microscopic organisms which contain no chlorophyll and are either parasitic or saprophytic on other plants and animals.

G

genotype The genetic make-up of an organism, as opposed to its physical appearance (phenotype).

Geographic Information System (GIS) A computerised mapping system that can incorporate several layers of information such as location, topography and vegetation, for display and analysis.

germ plasm bank Plantation concerned with the conservation of hereditary genetic material.

gley Poorly drained heavy textured mineral soil. Grey or bluish grey in colour. Low elevation gleys provide high yielding Sitka spruce crops if properly cultivated.

glulam (glue laminated timber) An engineered wood building product for load-bearing structures, made of board lamellas. These are finger-jointed lengthwise and then glued together with parallel fibres to produce an elongated beam which can be straight or arced.

greenhouse gas emissions Greenhouse gases include carbon dioxide (CO_2), methane (CH_4) and nitrous oxide (N_2O), which are released or emitted into the atmosphere as a by-product of natural and industrial processes.

group system A shelterwood system of successive regeneration fellings involving the removal of groups of trees.

gymnosperms vascular plants whose leaves are simple and opposite and that reproduced by means of exposed seed or ovule whose seeds are enclosed by ovaries or fruits. There are four extant gymnospermous plants including Pinophyta or conifers featured in this publication. See also angiosperms and broadleaves.

H, I, J

habitat Any place or type of place where an organism or community of organisms normally lives and thrives.

heartwood The dead, inner wood of trees. Contains tannins and other substances that usually darken the wood's colour. See sapwood.

hardwood General term used to describe the timber of broadleaved trees (see also softwood) although some hardwoods are comparatively soft such as lime, alder and poplar.

humus That more or less stable fraction of the soil organic matter remaining after the major portions of added plant (and animal) residues have decomposed (see mull, moder and mor humus).

indigenous species Species which arrived and inhabited an area naturally, without human intervention.

indurated soil A soil which has strongly compacted material, which is low in organic matter. It normally occurs at depths of 30-75 cm, and extends for 30-50 cm or more.

iron pan A hard impervious layer formed largely from iron compounds, having been washed down from the upper soil horizons.

juvenile instability The leaning or overturning of young trees (2 to 5 year old) due to the fracture or weakness of the root or base of the stem just below ground level. Affected trees are often not uprooted, but continue to grow, attempting to regain an upright position and resulting in stems with basal sweep.

L, M

leader (or leading shoot) The main or leading shoot of a tree.

lichen A symbiotic association of fungus and alga.

light-demander Refers to tree species that will not thrive in the shade. Most species are light-demanding including Sitka spruce, lodgepole pine, Japanese larch, oak and ash. See also shade-bearer.

lop and top Woody debris from thinning or felling operations. Also known as slash or brash.

marginal intensity An intensity of thinning which in terms of annual rate of volume removal is 70% of the MMAI, i.e. 70% of the yield class.

mast year A year in which seed is produced in exceptionally large quantities. Refers now to all species, but mainly associated with beech and oak which produce 'mast years' at intervals of 5-15 years in Ireland. The quantity of seed produced is determined by the amount of sunshine in the previous year's growth.

mature wood Also known as adult wood, is produced after the juvenile stage (in the case of Sitka spruce after 14 to 18 years). Mature wood has good strength qualities and is desirable for products that need strength and stability.

maximum mean annual volume increment (MMAI) The maximum average annual volume increment for the crop expressed in m^3/ha.

medium density fibreboard (MDF) Reconstituted panel board of medium density manufactured in Ireland by Coillte Medite from spruce and pine fibres which are bonded together with synthetic resins.

monoculture A stand or crop in which only one species or cultivar is present or largely dominates.

monoecious Relates to organisms in which separate male and female reproductive organs occur on the same individual. In the case of trees and some flowering plants, it applies to male and female reproductive organs in separate flowers on the same plant.

moder humus Intermediate between 'mor humus' and 'mull humus'. The current litter layer overlies partly decomposed material which is not matted as in 'mor humus'.

mor humus Raw humus; type of forest humus layer of unincorporated organic material; usually distinct from the mineral soil. Comprises current litter layer overlying a matted layer of partly decomposed material.

mull humus A humus-rich layer of forested soils consisting of mixed organic and mineral matter.

multi-functional forestry The objectives of multi-functional forest management are to maximise the social, economic and environmental – wood and non-wood - benefits.

mycorrhizae Beneficial soil fungi associated with tree roots.

N,O

National Inventory Detailed listing of standing wood volume in the national forest estate. Results are usually achieved by ground survey and remote sensing.

Natural Heritage Area (NHA) Area of nature conservation interest listed and mapped by the National Parks and Wildlife Service and which receive statutory designation. Includes areas previously listed as Areas of Scientific Interest (ASI)

natural regeneration The regeneration of a crop through seeds from mother trees on the ground or in the vicinity.

needle cast Defoliation of conifers as a result of disease or climate.

nurse species Species which enable more delicate or more site-demanding species to grow satisfactorily on what would otherwise be considered unsuitable sites.

oriented strand board (OSB) Reconstituted wood panelboard manufactured from pulpwood logs by bonding peeled wood strands which are arranged in layers at right angles to one another to provide strength, OSB is manufactured in Ireland by the Coillte owned company, SmartPly, at Waterford Port.

P

panel boards Wood products manufactured from wood chips and residues. Includes fibreboard, plywood, medium density fibreboard (MDF) and oriented strand board (OSB).

pathogens Organisms capable of causing disease.

peats Divided broadly into blanket peats – mainly in the west and mountain regions in Ireland – and the raised bogs in the midlands, derived from the accumulation of partially decomposed plants.

pelosol Fine textured clay or soil.

pH A value on a scale of 0-14 that gives a measure of the acidity or alkalinity of a soil. A neutral soil has a pH around 7 (range 6 to 8). Acidic soils have pH values of less than 7 but greater than 3 and alkaline soils have values greater than 7 but less than 9. The lower the pH the more acidic is the soil; the higher the pH the more alkaline.

phloem A tissue comprising various types of cells which transports dissolved organic and inorganic materials in vascular plants. See also xylem.**pioneer species** Generally, refers to a new plant or tree stage of development within a cyclical pattern of vegetation change or species that colonise a physical environment sequentially until climax species emerge and a final equilibrium ecosystem is achieved. See also climax species.

Programme for the Endorsement of Forest Certification (PEFC) An international accredited body which provides certification for forests and woodlands. PEFC provides a framework through which national and regional certification schemes can integrate internationally agreed criteria with their local circumstances. Formerly known as Pan European Forest Certification.

protected habitats or species Areas and organisms protected by the EU Birds and Habitats Directives.

plus trees The best trees or best stands which have been selected for tree breeding purposes. Plus trees – and surrounding trees – ideally should be free of ring shake and spiral bark patterns. They generally have large stem diameter, good vigour, straight main stem, low taper and an absence of low branching and forking.

provenance The location of trees from which seed or cuttings are collected.

pruning The removal of branches to produce knot-free timber.

public good forestry Refers to non-commercial forestry that benefits in areas such as climate change mitigation, recreation and managing unplanted, biodiverse areas. There is an expectation that public good forestry should be funded by the State.

pulpwood Small logs with a top diameter between 7 and 14cm. They are too small for economic sawing and are used in Ireland as the raw material for energy and panel board material.

R

reconstitution Replanting damaged or failed plantations.

Red Data Book Publication, which lists Ireland's rare and endangered living species at national and local level.

reforestation Natural or artificial restocking/regeneration. In Ireland usually refers to artificial restocking after the previous crop is clearfelled which forest owners are legally obliged to replant.

rideline Strip of land left unplanted to form a permanent internal forest boundary to facilitate planning, measurement and access. Rideline can coincide with existing boundaries such as streams and roads.

riparian Associated with river banks (from the Latin 'ripa' meaning river bank).

riparian vegetation Plant communities growing on river or stream banks. The flora on riparian zones can include grassland, flowering herbs, woodland, scrubland and wetland vegetation.

riparian zones (or areas) Refer to the land on each side of rivers or streams. Serve important functions including conservation (both fisheries and immediate environment), water purification, flood control, erosion prevention and landscape enhancement.

rotation The period of years required to establish and grow timber crops to a specified condition of maturity at which stage the crop is removed and reforested or in the case of continuous cover forestry or shelterwood system the understorey species forms the new dominant crop.

S

sapwood The living, outermost portion of a woody stem that transports water and minerals to the tree crown.

sawlog Logs, usually of at least 14cm top diameter, which are intended for conversion in a sawmill.

scrub Non-commercial unmanaged forest area.

sedimentation The process whereby soil particles are transported by surface water flow (water flowing above ground) into aquatic zones.

shade-bearers Understorey species (not always grown in the understorey) which are adapted to growing with limited access to light.

shelterwood system Silviculture system employing natural regeneration under partial shade of the mature crop.

silviculture The science of forest establishment, maintenance and management.

silviculture system The management process by which forest crops are tended, harvested and replaced.

Sites and Monuments Records (SMRs) Archaeological sites and monuments listed and mapped by the National Monuments Service of the Office of Public Works.

Society of Irish Foresters an all-Ireland organisation, founded in 1942 whose main aims are to spread knowledge of forestry and to improve professional standards in Irish silvicultural practice.

softwood General term used for the timber of coniferous trees although the timber of some coniferous species such as yew are much harder than some broadleaves i.e. lime and alder.

soil reaction The degree of soil acidity or alkalinity, usually expressed as a pH volume. Thus, acid soils would be described as having an acid reaction.

Special Area of Conservation (SAC) Area of significance for the conservation of special habitats which have been designated under the EU Council Directive 92/43/EEC on the conservation of natural habitats and of wild fauna and flora (commonly referred to as the Habitats Directive).

Special Protection Area (SPA) Area of significance for the conservation of special habitats which are important for birds and have been designated under the EU Council Directive 79/409/EEC on the conservation of wild birds (commonly referred to as the Birds Directive).

sprouting Beginning to grow; giving off shoots or buds.

stakeholder Individual or group which has an active involvement in an issue of public interest.

stand Refers to a forest stand, an aggregation of trees occupying a specific area sufficiently uniform in species composition, age distribution and condition as to be distinguishable from surrounding forests.

stocking Usually refers to the number of trees per hectare or in a stand. It could also refer to basal area, or volume per hectare/stand. It is generally used as a comparative expression (well stocked, poorly stocked or overstocked) and optimum stocking will depend on the age of stand and species.

sub-dominant trees These trees are not in the upper crown, but their leaders still have free access to light.

suckers New shoots produced from the base or under- ground roots of an established plant.

suppressed trees Trees whose leaders have no direct access to light and stand beneath the crowns of adjacent dominant, co-dominant and sub-dominant trees.

sustainable forest management (SFM) The stewardship and use of forests and forest land in a way and at a rate that maintains their biodiversity, productivity, regeneration capacity, vitality and their potential to fulfil, now and in the future, relevant ecological, economic and social functions, at local, national and global levels and does not cause damage to other ecosystems. (From the Ministerial Conference on the Protection of Forests in Europe, Helsinki 1993).

sustained yield The regular continuous supply of the desired produce to the full capacity of the forest.

T, U

taproot Main or central descending root.

tending Relates mainly to broadleaves. Involves the removal of poorly formed or defective trees that adversely affect the growth and quality of surrounding trees.

terminal height The height by which it is predicted that 40% of a stand will be windblown. It can be as low as 20m in some exposed sites.

thicket (or thicket stage) The time when the branches of trees meet and suppress ground vegetation. In the case of fast growing-spruce crops, thicket stage is reached as early as year seven.

thinning The regular removal of trees – beginning with immature and poorly formed stems – to improve growth and quality of the remaining trees.

thinning cycle The interval in years between thinnings.

top diameter The overbark diameter at the top end of any log.

top height The average height of the 100 trees of largest diameter per hectare.

top-dying This is a phenomenon causing crown discolouration and tree mortality in Norway or Serbian spruce. The exact causes of the phenomenon are not known but the condition seems to be related to the inability of the species to respond quickly enough to a sudden increase in water demand by the needles.

understorey Any plants or shrubs growing under a tree canopy.

unenclosed land Unfenced land that shows no evidence of having been improved and enclosed by man-made boundaries for agricultural use other than extensive grazing.

W, X, Y

windfirm Descriptive of trees and plantations that, because of species, soil or relative exposure, are unlikely to suffer windthrow.

windthrow Uprooting or breakage of trees caused by strong winds – also known as windblow.

xylem A plant vascular tissue consisting of various types of cells, which transport water and dissolved minerals from the roots to the rest of the tree including the leaves. Xylem contains parenchyma, a tissue that makes up the long woody fibres that support the tree. Secondary xylem provides the tree's dark rings, which determines its age as well as other dendrological information. See also phloem.

yield class A classification of rate of growth in terms of the potential maximum mean annual increment per hectare (m^3/ha/annum). Calculated as the total volume of wood produced, divided by the age at which maximum mean annual increment (MMAI) is achieved

INDEX

Page references to images are in **bold**.

A2 Architects 236
Abbeyleix, Co. Laois 30
acacia, two-thorned 40
Accoya 154
acoustics see under wood
Adare Manor, Co. Limerick 40, **41**
afforestation 18, 21–2, 39–44, 66, 69–72, 74, 77, 85, 100, 105, 112–14, 144, 187, 195, 223, 229
    future 255–60, 266
    private 42–3, 78, 80, 97, 191, 249
    in Scotland 129
    State 97, 192, 249
    State–private 190, 250
agriculture 17–18, 27, 42–4, 49, 56, 66, 72, 90, 100see also beef farming; carbon farming; dairy farming; forests; marginal land; tenant farmers
agroforestry 27, 57, 70, 79, 109–10, 191, 196–7, 220, 263 see also farm forestry
aisling (poems) 38
AK Ilen 172
alder, common (black) 32, 82, 84–6, 109, 213–14, **213, 214**
Alder Architects 182
Allaby, Michael 19, 257
Amazon rainforest 31, 50–2, 65
Amberla, Tomin 125
Ammannati, Bartolomeo 139
Andrew, Leslie 219
Anglo-Saxon 21
Annestown, Co. Waterford **107**
Anne Valley, Co. Waterford 106–7, **107**
apple 32, 109
aquatic ecosystems 104
arbutus 32
ARCEN Architects (Davagh Dark Sky Observatory) **243**
Arctic 52
Ardennes Forest, Belgium 103
Areas of Biodiversity Enhancement (ABEs) 80, **98**, 100
ash 30, 32, 36, 39, 55, 82–4, 86, 93, 109, 161, 164–7, 174, 183, 196, 206–10
ash dieback disease 43, 55, 87, 144, 161, 192, 196, **207**, 210
aspen 32
Athy Foundry 171
Attenborough, David 15–17, 57, 59, 63, 67, 70, 249, 256
Austrian Forest Programme 103

Avoca Forest, Co. Wicklow **198**
Avondale estate, Co. Wicklow 41, **44–5**, 190, 195, 217, 228, 238, **240**
    Beyond the Trees 137, 178–9, **178, 179**, 182, **231**

b210 architects 182
Bail, Murray 249
balloon-frame houses 124
Ballycurry forest, Co. Wicklow 262–3
Ballyhoura Mountains 94
Ballyogan Maintenance Centre, Co. Dublin 159, **160**
Ballyroan Library, Dublin 139, **140**
Bardic tradition 37–9
bat habitats 93
Battle of Kinsale 37
Battle of Moira 201
beech 23, 39, 82, 84, 86, 216–19, **217, 218, 219**
beef farming 72–3, 249, 255
Belleek Wood, Co. Mayo **70**
Białowieza Forest, Poland/Belarus 22–3
Big Red installation (London Festival of Architecture) **156–7**
biochar 134
BioClass system 80
biodiversity 17, 22–3, 57, 59–61,63, 66, 69, 80–100, 103, 107, 129, 191, 193, 196, 199, 203, 205, 210, 225, 228–31, 248, 250–1, 256, 261–2
biodiversity corridors 87 see also wildlife corridors
bioeconomy 16–17, 65, 68, 125, 150, 194, 250
BIOFOREST project 96
biofuels 131–3, 152, 193
biomaterials 131
birch 32, 82, 85–6, 107, 183, 211
    downy birch 32, 84, 211, **212**
    silver birch 32, 84, 211
bird habitats 96
bird species 80–1, 93–100
Birds Directive 99
BirdWatch Ireland 93, 96–7
Biscay, Bay of 105
Black Forest, Germany 77
blackthorn 32, 108
bluebells **86**, 201
BM Trada 134
board mills 147, 149, 153–4
boat-building 31, 118, 172, 202, 226,

243–4
boat restoration 172–4
bog roads 118, **118, 119**
bogs 21, 30, 34, 42, 80, 90, 102, 118, 190, 202, 205, 234–5, 257
    blanket 77, 80, 96, 189, 192, 228, 259
    raised 80, 118, 189, 228
Boland, Donal 118
Boland, Eavan 15, 37, 219
Bolsonaro, Jair 65
Bord na Móna 118
boreal forests 19, 49, 107
Bourke, Brian **28**
Brackloon Woods, Co. Mayo 30
Branek, Leo 142–3
Brazil 50–1see also Amazon rainforest
Breen, Billy 221
Brehon Laws 201–2
    Bretha Comaithchesca 32, 205–6, 211, 213, 215
Brendan Voyage currach **177**, 177–8
bridge-building 31, 116, 118, 180–2
broadleaves 21–2, 39, 43, 83–4, 86, 106–7, 109, 189, 192–3, 196, 251–2, 257
Brock Commons Tallwood House, Vancouver 135–7
bryophytes 80–1, 87
Bucholz McEvoy Architects 159
building see construction, wood in; construction industry; timber frame buildings
building regulations 158
Bulfin, Mary **108**
Burke, Edmund 75, 268
Butterfly Building (RBA and CS)182, **183**
buzzards 91–2, 94, **94**
Byrne, Liam 231, **232–33**, 262–3
Byrne, Paul 273

Caledonian Forest, Scotland 78
Calgary Central Library 139
Caman Wood, Co. Derry 239
carbon balance 66, 158
carbon code 74–5
carbon cycle 63, 67–9
carbon dioxide 67, 147
carbon displacement 16–18
carbon emissions 135, 158
carbon farming 74–5
carbon neutrality 16, 70, 109–10, 114, 151, 174, 252, 255–6, 266

carbon sequestration 15–16, 18, 66–8, 107, 130–1, 174, 194, 234, 251, 255–6
carbon sinks 66, 134, 193
carbon storage 16, 50, 125, **126–7**, 131, 134, 194
carbon substitution 17
carbon values 75
Carey, Michael 39, 91
Carney, James 31
Carrickfergus Castle, Co. Antrim 122, **123**, 174–5, **176**, 178, 182
Carson, Ciaran 34
Castle Caldwell, Co. Fermanagh 40, **41**
Castledaly Forest, Co. Galway **43**
cattle ranching 49
cedar, western red 190, 243–4, **243**, **244**, **245**
cellulose 129, 131–2
Center Parcs Ireland, Co. Longford 159, **161**
Chamshama, S.A.O. 100–1
Chang, Pao-Chi 119–20
charcoal 36, 128, 200
Chartres Cathedral, France 120
cherry 32, 82, 109, 214–16, **215**, **216**
Cherrywood Timber Canopy, Dublin **137**
chestnut 183
    horse chestnut 221
    sweet chestnut 39, 82, 86, 108, 221–3, **221**, **223**
Chiang Mai, Thailand **50**
churches, wooden 119
"Cill Chais" 30, 38
Clacton Spear 117, **117**
Clarendon Street Priory, Dublin 167, **168–9**
Clark, Kenneth 120
clearcutting 22, 260–3
Climate Action Plan 75, 78, 106, 114, 115, 174, 196, 234, 250, 266
climate change 15, 50, 56, 66–9, 90, 130–1, 197
    medieval 36
climate change mitigation 16–17, 67, 69, 146, 193–4, 252
Clonad Wood, Co. Offaly **62**, **87**
Cloragh forest, Co. Wicklow 231, **232-33**, **262**
Coen, Patrick 31
COFORD 70, 79, 112–13, 150, 214, 230, 266
Cognisense 43
Coillte 13, 17, **27**, 42, 63, 70, 80, 147, 150, 154, 191, 196, 250
    headquarters, Newtownmount-kennedy 159, **162–3**
Colbert, Jean-Baptiste 39
Coles, J.M. 116
Collins, Kevin 87, 201
commercial forests 13, 16, 27, 42, 57, 97–9, 104, 110, 114, 229, 255, 259
Confor 129
concert halls 138, 142–4
conifers 21–2, 42, 83–4, 86, 96, 106–7, 110, 161, 188–9, 196, 251–2, 257
Constantine, Basilica of, Trier 119
construction, wood in 117–25, 134–48, 155–9, 234–5, 256 see also timber frame buildings

construction industry 158, 179
Contemplation Space, Mater Hospital **167**
continuous cover forestry (CCF) 22–3, 27, 110, 196, 229–30, **231–2**, 262–3
Convention on Biological Diversity 80
Conversation Piece (Ruumiringlus and Hannigan Cooke) 183, **183**
Conway, Marina 85, **211**
Coolbaun, Co. Tipperary 211, **212**
Cooper, Robert T. 41
COP26 69, 130
Corlea Trackway, Co. Longford 118, **118**, 119
Coronation Plantation, Co. Wicklow **8–9**, **187**, 189–90, 258, 260
Covid-19 13–14, 17, 235
crafts see wood crafts
Crawford, Hugh 41
Creatomus Solutions (CS) 182–3
Cross, John 87, 201
crossbills 94
cross-laminated timber (CLT) 114, 133, 135, 137, 146–7, 155, 158–9, 159,225, 238
Cruagh Forest, Co. Dublin **242**
Cupforma Natura Solo 132, **133**
Curragh woodland, Co. Cork **86**
Curraghmorc Estate, Co. Waterford 228–9
Curtis, T.G.F. 81
Curtis, Tom 81
Cusack, Michael 206
Cyclone Friederike (2018) 53
Cygnum Building Offsite 147
cypress, Monterey 228

dairy farming 72–3, 107, 249, 255
Daltun, Eoghan 108
Dante Alighieri 20
D'Arcy, Gordon 99
Davagh Dark Sky Observatory, Co. Tyrone (ARCEN Architects) **243**
de Barra, Fionan 172
de Gaulle, Charles 11
deadwood **61**, 126
decarbonisation 16, 66, 114, 131, 145–6, 158, 179, 197, 234, 251
deer 92–3
    fallow deer 90, 92
    red deer 90, 92
    sika deer 23, 90, 92
deforestation 15–18, 66–7, 128–30, 257
    in Europe 52–5
    global 47–55, **48**, **49**, **52**
    in Ireland 21, 29–31, 37–40, 55–6
    tropical/sub-tropical 49, 130
Dendroctonus micans (great spruce bark beetle) 55
Denmark, energy sources in 152
Department of Agriculture, Food and the Marine (DAFM) 78, 250–2, 266
    Forestry Standards Manual 60
    Forestry and Water Quality Guidelines 103
Department of Agriculture and Technical Instruction (DATI) 41
Derrycunihy woodland, Co. Kerry **200**

desertification 15, 57
Design and Crafts Council of Ireland (DCCI) 164
Devil's Glen, Co. Wicklow 14, 20, **26**, 27, **87**, 192, **218**, 247
dlr LexIcon Library, Dún Laoghaire 139, 159
Doré, Gustav **20**
Douglas, David 238
Dowding, Paul 107
Dromkeen Wood, Co. Cork **82**
drought 13, 50, 53, 55, 100, 102, 130
Dublin Dental Hospital 171, **171**
Dublin Society 39–42, 190, 216, 223 see also Royal Dublin Society (RDS)
Dún Laoghaire-Rathdown County Council (dlr) 139, 159
Dunanore Wood, Co. Wexford **204**
dunlin 97, **97**
Dürer, Albrecht **38**
Dutch elm disease 37, 55, 87
Dyfllin Viking Ship 172

eagle
    golden eagle 90
    white-tailed eagle 90
Early Christian period 34–6
Earth Summit, Rio (1992) 130
egret
    cattle 96
    great white 96
    little 96
earthquakes 122, 135
ecosystems 67, 69, 80, 190 see also aquatic ecosystems; forest ecosystems
Edward, John 105
eight-toothed spruce bark beetle (*Ips typographus*) 55
elder 32
Electricity Supply Board (ESB) 11
Ellison, David 106
elm 32, 37, 82, 87, 184
    English elm 39, 55
    wych elm 39, 55
Elwes, John Henry 229
Enigmum VI and VII (Joseph Walsh Studios) **166**
Environmental Protection Agency (EPA) 72
engineered wood (mass wood) 17, 114, 122, 125, 135–6, 138, 155–60, 194
Ennos, Roland 234
Estonia 182–3
European Bank for Reconstruction and Development 155
European Commission 105
European Deforestation Regulation (EUDR). 53
European Economic Community (EEC) 11, 249
European Forest Institute (EFI) 16, 53
European Union (EU) 78
    Forest Law Enforcement, Gover ance and Trade (FLEGT) Action Plan 53
    Forestry Strategy 75
    Horizon 2020 57
evapotranspiration 100–2, **101**
Evelyn, John 39

Factor, Chaïm **220**
Fahy, Orla 93
Fallon, Peter 260
famine 30, 56, 172 see also Great Famine
farm forestry 73, 78, 108–10, 188, 191, 193, 214, 249, 253–5 see also agroforestry
Farm to Fork strategy 75
farming see agriculture; beef farming; carbon farming; dairy farming; tenant farmers
Farrelly, Niall 78, **79**, 259
fauna 90–100
    extinct 37, 90–1
    reintroduction 81, 90
Federal Forestry Agency (Russia) 52
fencing 118
fertilisers 85, 131, 134, 255
Finland 107, 143
fir 82
    Chinese fir 122
    Douglas fir 11, 14, 42, 82, 190, 196, 228, 238–40, **238**, **239**, **240**
    silver fir 23, 39
Fitzgerald, John 256
Fitzpatrick, H.M. 29, 223, 226, 240, 242–4
Fleming, Michael 99–100, 254, 259
Flight of the Earls 37
flood plain management 50
flood prevention 50, 104–6
flora 80–1, 86–7, 201, 210
    diversity 87, 90
fly agaric **87**
Fogong Temple, China 135, **136**
food security 17, 49, 56–63, 70
foraging 87, 107–8
Forbes, A.C. 223
Forbidden City, Beijing 122
forest
    definition 19–21, 27
    in Irish 20
forest canopies 80, 100
forest carbon cycle **68**, 69
forest certification 53, 61–3
forest degradation 49
forest ecosystems 11, 13, 15–16, 19, 22–3, 39, 59–60, 63, 65–6, 80–1, 83, 103–4, 106–8, 130, 199, 201–2, 231, 234, 248, 251, 262–3
Forest Environment Protection Scheme 201, 230
forest environmental guidelines 60
forest expansion 47
Forest Industries Ireland (FII) 65, 146
forest management 83 see also sustainable forest management
forest restoration 56 see also afforestation; reforestation
Forest Service 44, 103, 195, 201, 230, 267
forest soils 103
Forest Stewardship Council (FSC) 63
Forest Strategy (Ireland) 114, 196, 250–1
foresters 41, **44**
forestry 11, 13–16, 21–2, 27, 29, 40–1, 47, 69, 76–9 see also agroforestry; continuous cover forestry; farm forestry;

plantations
    and agriculture 70–5, **71**, **72**, 78 see also agroforestry; farm forestry
    attitudes to 66, 76, 250–1
    benefits 16–18, 56, 65–7, 102
    employment in 13, 60, 65, 69, 76, 110–14
    multipurpose 45, 49, 60–3, 66, 191, 193–4, 248
    nurseries 112
    and the rural economy 110–14
    State 41–5, 96–7
Forestry Act (2014) 19
Forestry Programme (2023–27) 248
forestry rotation 22, 69, 74–5, 77, 85, 127, 193, 196, 201, 214, 219, 221, 229–30, 239, 260–2
forestry systems 260–3
forests see also boreal forests; commercial forests; temperate forests; tropical forests
    burning, deliberate 17, 31, 50–2, 63, 65–7 see also wildfires
    coastal 105
    European 76
    in folklore 32–4
    location 257–60
    mixed 80, 196
    open areas in 87
    overcutting 13, 16, 29, 50, 59, 60, 67, 128
    overstorey 205
    planted see plantations
    in poetry: 32–4; Early Christian 34–6
    primeval 30–4, 80
    public goods 49–50, 66, 69, 74–5, 106–10, 256
    recreational use 13–14, 17, 60, 66, 69–70, 106–8, 110, 191, 251, 255
    State-owned 250
    threats to 53–5
    and tourism 106
    types 196
    understorey 22–3, **82**, 84–6, 199, 226, 263
    upper storey 263
    and water 100–5
Forisk Consulting 133
fossil-based materials, displacing 16–17, 56, 63, 66–8, 114, 125, 128–31, 133–5, 138, 145, 154–5, 174, 184, 194, 196, 251
fossil fuels 16, 129, 131–2, 134, 152–3, 250
    alternatives to, see biofuels
France 158
fungal disease 53, 55, 183, 192, 196, 210, 220, 244
fungi 80, 87, 107, 130
furniture 13, 16, 31, 45, 68, 113–14, 115–16, 125, 135, 144, 154, 161, 164–70, 179, 193, 199, 202, 206, 210–11, 215–17, 219–21

Gaelic culture, decline of 37–8
Gallagher, Gerhardt 78, **79**, 259, 267
gallowglasses **38**

garden furniture 239, 243
Gem Joinery 175
Georgius Agricola 65
Gilbert, Gillian 93
Glanmore, Co. Wicklow 14, 20, **192**
Glendalough Forest, Co. Wicklow 11, 30, 228, 238
    Lugduff **12**
    Poulanass (*Poll an Easa*) **10**
Glengarriff, Co. Cork 30
Glennon, Paddy 177
Glennon Brothers Sawmills 175, 177
global warming 13, 15, 49, 52–3, 55, 66–7, 234, 256
glulam (GLT) 122, 135, 137, 143, 159–60, 225, 231, 238
Gold Tulip (Emmet Kane) **170**
Goldsmith Street housing, Norwich 147–8, **148**
Goldsworthy, Andy 76
Goodall, Stuart 159
Gortian Warm Stage 223
Gougane Barra, Co. Cork 7
Graham, W.S. 217
Granhults Church, Sweden 119, **121**
grasshopper warbler 96
Great Famine 41, 66
great spruce bark beetle (Dendroctonus micans) 55
greenhouse gases (GHG) 17–18, 52, 56–7, 66–7, 69, 72, 74, 146, 151, 252, 254–5
Greenpeace 52
ground cover 27, 86, 200, 242
Guarneri, Giuseppe 142
Guterson, David 243
Guilfoyle, Michael 267

habitat loss 15, 60, 90, 99
Hall, Valerie 36
Hammond, Fred 171
Hannigan Cooke Architects (Conversation Piece) 183, **183**
hardboard 154
hardwoods 124–5, **127**, 160–4, 198–9
    tropical 125
harrier, hen 97, 99–100
Harrington, Rory 106–7
Harsta, Atto 128
Hart, Cyril 205–6
Harte, Annette 154
Hartnet, Michael 38
hawthorn 32, 34, 108
Hayden, Thomas J. 90, 93
Hayes, Samuel 40, 217
hazel 32, 34, 84, 108
Hazel Wood, Co. Sligo **69**
Heaney, Seamus 14, 20, 29–31, 34, 47, 176, 184, 202
hedgerows 32, 55, 77, 79, 83–4, 87, 90, 93, 97, 199, 207, 220, 256
Helsinki Central Library (Oodi) 139
hemicellulose 131–2
hemlock, western 23, 190, 246–8, **247**
Henry, Augustine 42–3, 65, 105, 190, 195–6, 229, 229
HoHo skyscraper, Vienna 137
Holder Mathias Architects 159
holly 27, 32

Holohan, Donn 182
Holzbrücke Rapperswil-Hurden, Lake Zürich 118
Horgan, Ted 214, 221, 239, 244
hoverflies 93
Hughes, Robert 124
humus 85
hunter-gatherers 80, 117
hurleys 206, 210
hydroxyethyl cellulose (HEC) 132
Hymenoscyphus fraxineus see ash die-back disease

Ice Age 81 see also Little Ice Age
Indonesia 52
insect habitats 93
Integrated Constructed Wetlands (ICWs) 106–7
International Tropical Timber Agreement 130
Inter-Party Government, first 42
invasive species 234–5
Ips typographus (eight-toothed spruce bark beetle) 55
Irish Bioenergy Association (IrBEA) 152
Irish Farmers Association (IFA) 73, 99, 254
Irish Forestry Society 41
Irish language 20, 32
Irish Rare Birds Committee 93
Irish Timber Frame Manufacturers Association (ITFMA) 147
Irish Timber Growers Association (ITGA) 96, 225
Irish tree alphabet 33
Irish Water 106
Iron Age 118
Irwin, Rachel 255
Irwin, Sandra 83, 85
Iwata, Yuki 70

Jodidio, Philip 115
John Sisk & Son (Holdings) Ltd 159
Joyce, James 41, 195, 206
Joyce, P. 234
Joyce, P.W. 36
juniper 32, 202

Kane, Emmet 170
    Gold Tulip **170**
Kavanagh, Patrick 185–6
Keithia thujina 244
Kelly, Fergus 32
Kilbeggan Distillery, Co. Westmeath 171, **171**
Killarney, Co. Kerry 23
Kilpatrick, C.S. 239
Kinsella, Thomas 29
kite, red 90, 94
Knut Klimmek Furniture, Mu Table **215**
Kordik, Hans 150–1
Korpella, Salla 107

Laffey, Micheál 85
laminated veneer lumber (LVL) 122, 135
Land Acts 41

Land Law (Ireland) Act (1881) 39
land use 76–9
Landes, Forest of, France 77, 105
landlords 39, 40–1
landscape 76–9
lapwing 96–7
L'Aquila concert hall, Italy 142, **143**
larch 84, 86, 183
    European larch 11, 39, 82, 84, 226–8, **226**, **227**
    hybrid larch 226–8
    Japanese larch 27, 42, 55, 82, 84, 240–3, **241**, **242**
    Xingan larch 135
Latin 21
Laurentian Library, Florence 139
Lawton, Colin 91
Leskinen, Petra 16
Levey, Michael 139
Lewis, Lesley 93
libraries 139–1
lichens 80, 87, 201, 219
lignin 129, 131–3
lime 82
Lindroth, Jan 131–2
linnet 96
Lintula, Kimmo 143
Little Ice Age 38–9 see also Ice Age
Lobb, William 243
logging, illegal 15, 17–18, 53
logrolling 119
Longfellow, Henry Wadsworth 20
Lowery, Martin 73
Lucas, A.T. 201
Lurgan Longboat/Canoe **30**, 31, 190, 200
Lusby, John 94, 99

Maam Valley, Co. Galway 85, 206, 260
MacBride, Seán 42
Mac Carthaigh, Muiris 266–7
McCarthy, John 267–8
McCarthy Keville O'Sullivan Ltd 103
McCauley, Kenny **211**
McCauley Wood Fuels, Co. Leitrim 152
McCullough Mulvin Architects 171
McGough, H.N. 81
Mc Loughlin, John 215
McManus, Damien 31
McMorrow Haulage & Firewood, Co. Leitrim 152
MacNeill, Máire 34, 36
Magh Rath, Battle of 20
Magnus del Busolo (Joseph Walsh Studios) 165, **164–5**
Magnus Celestii (Joseph Walsh Studios) 165, **165**, **208–9**
mammals 90–3
maple 23
    field maple 82
    Norway maple 82
"Marbhan and Guaire" 34, 108
marginal land 18, 41, 43, 70, **71**, 72–3, 75–6, 78, 192, 249, 254, 258–9
Marnell, Ferdia 93
Marvell, Andrew 128
Masonite Ireland 154
masonry 119–20, 175

mass wood see engineered wood
Matthews, Alan 73
Medieval Warm Period 36, 39
Medite Supply 154
medium-density fibreboard (MDF) 113, 154, 234
Menai Bridge, Wales 122
Menzies, Archibald 238
Meredith, Alan 165–7, 170
    Vinculum series, **167**, **170**
    oak vessels, **170**
merlin 97, 99
Mesolithic period 30, 117
methane 67
mice, wood 90
Michelangelo 139
Mitchell, Frank 32, 223
Mitchell-Jones, Tony 93
Mjøstårnet, Brumunddal, Norway 137, **138**
Model School, Inchicore **222**
Moffett, Marian 118
Monivea, Co. Galway 40
monoculture 13, 21–2, 199, 231
Moorepark, Co. Cork **108**
Morris, Locky, Polestar, Letterkenny **205**
Morris, Stephen 172
Moss, Richard 40
mosses 80, 87, 201, 235
mountain ash 23, 32, 84–5, 164, 206, 263
Muldoon, Paul 34, 108
Mullaghmeen Forest, Co. Westmeath 217
Munro, Alice 19, 27, 217
Murphy, Gerard 34
Murphy, Gerry 90, 91, 96
mushrooms 224, 225
musical instruments 142
Mu Table (Knut Klimmek Furniture) **215**
mycorrhiza 87
Mylne, Alfred 173–4

Nanchan Temple, China **118**, 119
Naneen 171–4, **171**
nanmu 122
National Forestry Inventory (NFI) 80, 234
National Parks and Wildlife Service (NPWS) 13, 17, 42, 70, 90, 99
Native Tree Area (NTA) scheme **108**, 109
Native Woodland (Conservation) Scheme 44, 57, 85, 196, 201, 213–14, 230
Natural Resources Canada 132
Neff, Jenny 81
NeighbourWood Scheme 44, 70
Nelson, Charles 211
Neolithic period 30, 119
Netherlands 158
New York Public Library, Wallach Division 139
Newcastlewest, Co. Limerick **70**
Newgrange passage tomb, Co. Meath 119

Ní Dhubháin, Áine 255
nightjars 94
Nikken Sekkei 138
Nisbet, Tom 105–6
Nisbet, John 41
nitrous oxide 67
None so Hardy Nurseries **110–13**
Northern Ireland Forest Service (NIFS) 17, 63, 70
Norway, 90
Notre-Dame Cathedral, Paris 120, 122
nuts 107–8
Nwonwu, F.O.C. 100–2

oak 11–13, 30–2, 36, 75, 82–3, 118, 122, 170, 174, 199–202, 206
    pedunculate oak 84, 200, 202
    sessile oak 84, 199–200, **200**, 202, **264–5**
Oak Vessels (Alan Meredith) **170**
O'Callaghan, C.J. 96–7
OCarroll, N. 231
O'Connor, Ciaran 148, 160, 167, 174
O'Connor, J. 137
O'Connor, Matt 143
O'Donnell + Tuomey Architects, Vessel 175–6, 177, 178, 182
O'Donoghue, Cathal 70, 255, 267
Ó Foghlú, Ray 259–60
Ogham 31–3
O'Halloran, John 96, 231
O'Keeffe, James George 34, 201, 206
Oliver, C.D. 129
Ophiostoma novo-ulmi 87
Ophiostoma ulmi 87
Ó Rathaille, Aogán 29, 38–9, 128
oriented strand board (OSB) 113, 154, **156**, 234
Orkney 117
osprey **90**
O'Sullivan, Aidan 118
O'Sullivan Beare, Don Philip 221
O'Toole, Des 146–7, 152
owl, long-eared **95**, 96

Palaeolithic period 30
pallets 234–5, **236–7**
palm oil 49
paper production 124–5
paperboard 125, 132
Parapox virus 91
Parco della Musica, Rome 142, **143**
Parnell, John 81
Passivhaus 147–8
Patterson, Gerry 13, **219**
pearl mussel, freshwater 79, 103–4, 259
peatlands 78 see also bogs
Peter Brett Associates 159
Peter Pere Architects 183
photosynthesis 68
Phytophthora cambivora 214
Phytophthora ramorum 55, 87, 226–8, 242
Piano, Renzo 142
pine 21, 30, 82, 84, 105
    lodgepole pine 21, 42, 84, 190, 228
    maritime pine 105
    Monterey pine 228
    Scots pine 11, 27, 32, 39, 82, 84–5,

107, 187–9, 196, 202–6
    swamp pine 40
    Weymouth pine 39–40
pine marten 91–2, **92**, 219
placenames, trees in 32, 36, 117, 201, 206–7, 211, 213, 215
planning laws 155
plantations, the 21
plantations 21–2, 57, 83, 85, 99–100
    new 125
plants see flora; vascular plants
plastics 132, 134
plover, golden 97
Polestar, Letterkenny (Locky Morris) **205**
pollution 103, 106, 220
poplar 110
    Athenian 40
population, global 56
Powerscourt Estate, Co. Wicklow 239
Pres-Lam 122
Pre-Stressed Timber Ltd 122
Programme for the Endorsement of Forest Certification (PEFC) 63
Project Ireland 2040 146
Project Woodland 251

Rackham, Oliver 19–20
Rain Bridge, Fujian Province, China **180–1**, 182, **182**
raptors 90
Raven, Co. Wexford 105
Ravna vala, Bosnia and Herzegovina **22**, 23
Raw Potential (Peter Pere and Workshop) 183
Raworth, Kate 183
red kite 90, 94
redstart, 94
redwood, coast 228
reforestation 16, 42–3, 60, 85, 94, 111–12, 128, 187–8, 195–6, 260–1
renewable energy 16, 66, 125, 150–4, **153**, 194, 196, 235–6, 256 see also wood energy
Restoration of Mountain Lands (*Restauration des Terrains en Montagne*), France 105
rewilding 80, 108
rhododendron (*Rhododendron ponticum*) 23, 55
Riches, Mikhail 148
Riordan, Maurice 34, 36
Ritchie, Hannah 132
River Basin Management Plan 103
river catchments 100–3, 106, 259
river habitats 104
rivers 102
Robert Bourke Architects (RBA) 182
Roche, Jenni 202–5, **202**
rock phosphate 85
Rockforest Lough, Co. Clare 202–3, **203**
Romunde, Richard 158
Rooney, Seán 90, 93
Roser, Max 132
Rotary Ireland 14
rowan see mountain ash
Royal Dublin Society (RDS) 252–3 see also Dublin Society

Rüdiger Lainer and Partner (RLP) 137
Ruumiringlus architects (*Conversation Piece*) 183, **183**
Ryan, Mary 43
Ryan, Michael 32, 223

St John's Wood, Co. Roscommon 30
Sally O'Keeffe 172, **173**, 174
Samuel Beckett Civic Campus, Co. Dublin 159
sand dunes 105
Sathre, R. 137
sawmills 11, 124, 147–50, 158–9, 165, 235
scaffolding, timber 119
Scandinavia 77 see also tree species
Schiamberg Group 133
"Scribe in the Woods" 34, **35–6**
Seán Harrington Architects 236
setback areas 86–7, 90, 104, 234, 248, 259
Severin, Tim 177
Shakespeare, William 20
Shannon, River 118
Sheehan's Sawmills, Co. Tipperary 171
shipbuilding 36, 116, 124, 172, 200 see also boat-building; boat restoration
shrews 90
Sibelius, Jean 143
Sibelius Hall, Lahti, Finland 143, **144**
Siberia 52
siskins 94
Short, Ian 255
Smith, Louis 107
Smyth, William J. 40
snipe 97
Society of Irish Foresters (SIF) 42, 74, 256
Södra 133
softwoods 124–5, **126**, 144–5, 154, 160–4, 198–9, 234
soil conservation 50
soil erosion, prevention 105–6, **212**
Somers, Michael, **212**
soya beans 49
Spain, forests in 103
Special Protection Areas (SPAs) 99–100
spindle tree 32
Spooner, Paul 172, 174
spruce 82–3, 158
    Norway spruce 23, 27, 39, **54**, 84, 107–8, 183, 196–7, 223–5, **223**, **224**
    Serbian spruce 197
    Sitka spruce 13, 21, 41, 82, 84, 93, 96, 105, 108, 155, 159, 175, 183, 190, 192, 196, 223–5, 228–38, **228**, **230–1**, **233–4**, **235**, 239
spruce bark beetle 55
squirrel, 90
    grey **91**, 90, 94, 213, 219, 220, 234
    red, **91**, 92, 219, 234
Stanbury, Andrew 93
Standish, Sarah **108**
Stapleton, Myles 172
Staunton, Pat 14
Stewart, Duncan 159
Stone Age 30
stone structures see masonry
Stora Enso 16, 132–3

Storm Darwin (2014) 55–6
Storm Ophelia (2017) 122
storms 41, 53, 55, 105, 195, 261
Stradivari, Antonio 142
Styles, David 70
Suhonen, Timo 125
Suibhne (Sweeney) mac
Colmáin/Suibhne Geilt 20, 29, 34, 201, 206
Sumitomo Forestry 138
Sundström, Erik 229
Sundström, Karl-Henrik 16, 115, 131–2
Support Scheme for Renewable Heat (SSRH) 152
sustainable forest management (SFM) 15–18, 22–3, 44–5, 50, 53, 56, 59–63, 67, 70, 103, 105, 128–30, 152, 174
sustained yield 59–60
Sweden 16–17, 73
Swenson, Alfred 119–20
Swift, Jonathan 39
sycamore 23, 82, 86, 110, **219–20**, 219, 220

taiga 19
Taylor, Elysia, rocking chair **217**
Teagasc 210, **212**, 214, 255
Technological University Dublin 14
temperate forests 22, 49, 51
tenant farmers 40–1
Tikka, Hannu 143
timber 53, 59–60, 73 see also wood
timber crops 109
timber clearances 117
timber exports 149
timber frame buildings 114, 124, 126, 128, 134, 137, 145–7, **146**, **147**, **148**, 154, 194, 225, 234–5
timber products 113–14
Timpany, Scott 117
Tiseo, Ian 132
Tittensor, Ruth 229
Tokyo Opera City Concert Hall 142
Tomies Wood, Co. Kerry 30
Toner, Gergory 32
tories 37
transmission poles 11
Tree Register of Ireland 223
tree rings 125
tree species 84 see also placenames, trees in
    diversity 192
    European/naturalised 83, 189, 216–28
    exotic 197, 199, 228–48
    grant-aided 197
    introduced 81–3, 195–7
    Irish names 205–6, 214
    for Irish conditions 195–248
    medicinal uses 108
    native 81–2, 189, 196, 199–215
    North American 83, 187–90
    nurse species 211, 214, 226, 239, 242, 244, 246
    pioneer species 211, 226, 235
    Scandinavian 188
    selection 185–94, 255–6
Treet (The Tree) building, Bergen 137
Tricoya 154, **154**

Trinity College Dublin, Long Room Library 139, **140–1**
tropical forests 15, 22, 47, 51
Tudge, Colin 17, 115, 266

United Nations (UN)
    Environment Programme 52
    Food and Agriculture Organization (FAO) 49–51, 56, 106
    Forum on Forests 59
University of British Columbia 137
University of Canterbury, New Zealand 122, **122**
Uragh Wood, Co. Kerry 30
Urnes church, Norway 119

Vale of Clara, Co. Wicklow 23, **24–5**, 27, 30
van der Lugt, Pablo 128, 158
van Romunde, Richard 114
vascular plants 22, 80–1, 87, 90, 197
Veitch, John Gould 240
Venice Biannale (2012) 175
Verstraeten, W.W. 100
Vessel (O'Donnell + Tuomey Architects) 175–6, **177**, 178, 182
viburnum 55
Vienna 103
Vinculum series (Alan Meredith) 166–7, **167**, **170**
Virgil 187, 191, 260
vole, bank 90
Voll Arkitekter 137
Voluntary Partnership Agreements (VPAs) 53

walnut 39, 110
    black walnut 143
Walsh, Joseph **164–166**, 179, **208–209**
warbler, grasshopper 96
water, drinking 102–3, 106
water conservation 50
water cycle 100
water quality 102–6
Webb, David A. 81
West Telemark Museum, Norway 119, **120**
Wexford Harbour 105
Wexford Opera House 143, **145**
Whelan, Donal 94–6
whinchat 96
whitebeam 32
whitethorn see hawthorn
Wicklow, Co. 39
Wicklow Mountains 258
wildfires **46**, 50–2
wildlife corridors 83, 93, 234 see also biodiversity corridors
willow 27, 32, 84–5
Wilson, M.W. 97
windthrow 50, 55–6, 130, 183, 229
wolves 90
wood see also engineered wood; timber
    acoustic properties 138–43, 167
    aesthetic qualities 138
    benefits of 16
    constituents 131–2
    markets 198–9

recycled 16
uses of: 115–84, **126**, **127**, 202, 206, 210–11, 217, 219–21, 224, 243–4see also construction, wood in; furniture; timber-frame buildings
primeval 32, 117
wood (place), definition 20–1, 27
Wood Awards Ireland (WAI) 159, 165, 167, 171, 179
wood-based panels (WBPs) 135, 152–5
wood-based products 16
wood biomass 67–8, 150
wood chips 150
wood crafts 160–70
wood energy 16, 67, 150–2 see also renewable energy
Wood Fuel Quality Assurance (WFQA) scheme 152
wood-kernes 20, 37, **38**
wood production 56–7, 59–60
wood products see timber products
wood restoration 170–1
wood security 17, 56–63, 70, 160
wood supply, sustainable see sustainable forest management
wood turning 170
wood waste 131, 150, 152–3
Wood Works project 192
woodland 17, 22, 31–2, 34, 36–44, 57, 63, 65, 70, 75–87, 90–4, 96, 109, 161, 185, 219–21, 224, 248, 253, 257–8, 260, 263 see also forests
    conservation 85
    definition 19–21, 27
    in Irish 20
    mixed 196, 216–17
    native 11–13, 21, 103–4, 107, 110, 129, 189–90, 196, 199–203, 205, 210–16, 248
    opposition to 103–4
    riparian 105–6
    semi-natural 23, 30, 42, 85
    State-owned 250
    upland 105
woodland development 186–7, 191, 195–7
woodland improvement 111
Woodland Trust 225
"Woodland for Water" **104**
woodlarks 90
woodpeckers 94
    great spotted woodpecker 94, **96**
wood warbler, 94
Wordsworth, William 219
Workshop Architects 183
World Wildlife Fund (WWF) 80
Wormslev, E.C. 133

Yanagisawa, Takahiko 142
Yasaka Pagoda, Kyoto 122
Yeats, W.B. 41
yew 32, 116–17, 202
Young, Arthur 40